AF846477

Human Brain

Scientific and Psychological Factors

Editors

Emre Tokgoz

School of Engineering and Technology
State University of New York
Farmingdale, NY, USA

Sophia Marra

School of Medicine, Quinnipiac University
North Haven, CT, USA

Percin Karakol

Başakşehir Cam and Sakura City Hospital Plastic Reconstructive
and Aesthetic Surgery
Istanbul, Turkiye

CRC Press
Taylor & Francis Group
Boca Raton London New York

CRC Press is an imprint of the
Taylor & Francis Group, an **informa** business
A SCIENCE PUBLISHERS BOOK

Cover credit: Image reproduced by kind courtesy of the editors.

First edition published 2027
by CRC Press
2385 NW Executive Center Drive, Suite 320, Boca Raton FL 33431

and by CRC Press
4 Park Square, Milton Park, Abingdon, Oxon, OX14 4RN

© 2027 Emre Tokgoz, Sophia Marra, Percin Karakol

CRC Press is an imprint of Taylor & Francis Group, LLC

Reasonable efforts have been made to publish reliable data and information, but the author and publisher cannot assume responsibility for the validity of all materials or the consequences of their use. The authors and publishers have attempted to trace the copyright holders of all material reproduced in this publication and apologize to copyright holders if permission to publish in this form has not been obtained. If any copyright material has not been acknowledged please write and let us know so we may rectify in any future reprint.

Except as permitted under U.S. Copyright Law, no part of this book may be reprinted, reproduced, transmitted, or utilized in any form by any electronic, mechanical, or other means, now known or hereafter invented, including photocopying, microfilming, and recording, or in any information storage or retrieval system, without written permission from the publishers.

For permission to photocopy or use material electronically from this work, access www.copyright.com or contact the Copyright Clearance Center, Inc. (CCC), 222 Rosewood Drive, Danvers, MA 01923, 978-750-8400. For works that are not available on CCC please contact mpkbookspermissions@tandf.co.uk

For Product Safety Concerns and Information please contact our EU representative GPSR@taylorandfrancis.com. Taylor & Francis Verlag GmbH, Kaufingerstraße 24, 80331 München, Germany.

Trademark notice: Product or corporate names may be trademarks or registered trademarks and are used only for identification and explanation without intent to infringe.

Library of Congress Cataloging-in-Publication Data (applied for)

ISBN: 978-1-041-07453-3 (hbk)
ISBN: 978-1-041-08575-1 (pbk)
ISBN: 978-1-003-64604-4 (ebk)

DOI: 10.1201/9781003646044

Typeset in Times New Roman
by Prime Publishing Services

Preface

Human brain research has influenced many different aspects of life by helping us understand how people learn, behave, decide, get sick, and many other factors. Both psychological and physiological factors play a significant role in our lives, and such results of research are even incorporated into different areas of artificial systems' development. For instance, the notion of artificial neural networks is developed based on the logic of neural networks that rely on brain functioning and the associated development of digital networks and artificial intelligence solutions. While we need to acknowledge that human brain research is very extensive with no single book's ability to cover all concepts, this book relies on the published peer-reviewed relevant research articles and well-known healthcare related institutions' provided information on the following scientific, medical, and psychological aspects of the human brain:

* Chapter 1: Brain's role in learning
* Chapter 2: Brain's role in anger development and management
* Chapter 3: Brain's role in social aspects in life
* Chapter 4: Brain's role in mathematical comprehension, learning, and thinking
* Chapter 5: Brain's role in Alzheimer's disease
* Chapter 6: Brain's role in autism spectrum disorder
* Chapter 7: Brain's role in intellectual disability considerations
* Chapter 8: Brain's role in attention deficit and hyperactivity disorder
* Chapter 9: Brain's role in addiction
* Chapter 10: Brain's chemistry for common neurological disorders
* Chapter 11: Brain's role in attention control and anxiety management
* Chapter 12: Brain's role in beauty and body perception
* Chapter 13: Brain's role in gender reassignment decisions
* Chapter 14: Brain's role in limb amputation related considerations

Each chapter in the book is structured in an article format with the authors' contributions of each chapter are specified, containing a total of approximately

1100 article reviews. The amount of information provided in this book is extensive and can be utilized as a part of undergraduate degree or graduate degree courses.

We hope you will enjoy reading and learning about different aspects of the book!

Thank you,

Emre Tokgoz
Sophia Marra
Percin Karakol

Contents

Chapter 1

The Role of Brain on Learning

Emre Tokgoz,[1,*] *Sophia Marra*[2] and *Percin Karakol*[3]

1.1 Introduction

The human brain is the director of organs that function to learn and manage memory with a complex structure. It's partnering with other systems as a part of a greater planned learning and comprehension system including many organs and sub-systems of the body that researchers are still trying to understand. This complex system has many different ingredients in connection with learning. Researchers have focused on different aspects of learning and memory throughout the centuries with different factors included. In this work, some of these factors would be explained based on the theories that are developed and a variety of modeling schemes generated for understanding how the brain functions in collaboration with other organs and systems. Such factors to be covered include the mix of the following:

- Learning theories
- Memory
- Attention
- Nervous system
- Neuropsychology
- Neuroscience

[1] School of Engineering and Technology, State University of New York, Farmingdale, NY, 11735, USA.
[2] School of Medicine, Quinnipiac University, North Haven, CT, 06473, USA.
[3] Başakşehir Cam and Sakura City Hospital Plastic Reconstructive and Aesthetic Surgery, Istanbul, Turkiye.
* Corresponding author: tokgoze@farmingdale.edu

DOI: 10.1201/9781003646044-1

- Environmental factors
- Stress
- Anxiety
- Organs such as eyes, hands, and ears

Given the extensive nature of the brain and the associated systems and their subsystems taking place in the operations, this chapter attempts to explain a certain portion of the accomplished research towards how learning works by the brain in collaboration with other organs.

1.2 Memory Models, Theories, and Brain's Role

The brain has a dominant role in the foundations of the models and theories developed for memory models developed in the literature that was dominantly initiated after the 1950s. In its simplest form of explanation, the models and theories on how memory works to be explained in this section influenced other researchers in the literature not to further analyze these models and theories through further experiments but also to develop the modern world understanding of artificial intelligence and the associated neural networks and object recognition and detection. The models to be covered in this section include the following:

- Multicomponent Working Memory Model
- Selective Filter Theory
- Attenuation Theory
- Bottom-Up Theory of Perception
- Multistore Model of Memory
- Top-Down Theory of Processing
- Computational Theory

The details of the models and theories are left to the reader to review the corresponding references for further details.

1.2.1 Multicomponent Working Memory Model

The notion and terminology of working memory was introduced in the 1960s [8]. The initial focus of the memory was being a short term or long term. The multicomponent working memory model is the most popular working memory model type that is used by researchers with a focus on multiple components with ever changing memory structure [1–7]. The complexity of the cognitive functioning and ability with a continuum of ever-changing memory conditions create the multiple components to be considered in the memory model. In this model, the following three components are taking place [9, 10]:

- *Visuo-spatial scratch pad (inner eye)* – The working memory for visual-spatial portion that is designated for temporary storage and manipulation of the spatial and visual information.

- *Phonological loop* – Dedicated to the verbal portion of the working memory with an articulatory control system that serves to temporarily hold verbal information.
- *Central executive portion* – Control system in charge of the attentional control that is independent of modality and determined to have capacity limitation.

A fourth component called the Episodic buffer was included in the working memory model to represent the temporary storage system for integrating different sensory information with their modulation [11]. Overall, the central executive portion serves as the recalling, manipulating, and processing of the verbal and non-verbal information for many different applications that include but are not limited to writing, decision making, and problem solving. It is possible that this information could be registered to the long-term memory [12]. Long-term storage is determined to be not necessarily an extension of the short-term memory but rather highly related to the long-term memory [13].

The complexity of the memory spanning through multiple timescales and stages with the involvement of multiple brain regions makes it a complex process that researchers attempted to model parts of the entire system mathematically. The mathematical models developed for memory processes are either focused on the specific regions of the brain or meaningfully connected to a biological substrate [123–125]. An attempt to patch this gap through the use of a mathematical model to explain memory comprehensively through neuron interpretations was made in [122]. A well-known detailed biological analysis of ACT-R's declarative memory system is attempted to be used for providing such details. This framework outlines the formation, retrieval, merging, forgetting, and making mistakes in the memory's systematic design. This architecture of the brain relates to the memory and long-term memory model ACT-R [122]. In this model, the encoding process with the individual features f_1, f_2, ..., f_N are derived from different cortical regions that are bounded together into a new memory in a region within hippocampus. The retrieval process is based on the integrated work of constraints that are held in the lateral PFC, and similarity across memories and the spreading activation from contextual cues q_1, q_2, ..., q_N, and the competing memories' degree of activation that led to the selection of a recalled memory. Information communication back to the cortical regions occurs with the retrieved information from the recalled features with additional features attained from the competing memories.

1.2.2 Selective Filter Theory

The idea behind the Selective Filter Theory by Broadbent (1958) is centered around the dismissal of processing unattended stimuli that could exceed the basic physical properties. The limitations are centered around the following:

- Central executive that has a limited capacity and serves as the control system with a modality free structure.
- *Phonological store* – Inner ear with speech-based storage system

- *Phonological loop* – Articulatory control system
- *Visuo-spatial scratch pad (inner eye)* – Temporary storage and manipulation of visual and spatial information

Numerous findings attained later motivated researchers to reject this theory; however, additional evidence later indicated strong reasons for beliefs to doubt that the associated irrelevant stimuli are unattended. Experimental work on five cases by using priming paradigm in [14] resulted in no priming effect production when all the conditions were set for the irrelevant stimuli to be not attended; It was concluded by the authors that there is no reason for identification of unattended stimuli based on the five experiments.

In relation to the anatomic structure and selective filtering theory, the role of neural networks has a significant impact on the outcomes. A demonstration of conscious and preconscious processing is demonstrated in [62]; The preconscious processing with the activation pattern of the brain is confined to sensory-motor areas and does not reach higher parieto-frontal areas.

1.2.3 Attenuation Theory

Attenuations theory is proposed by Treisman in 1960 by assuming that the unattended input diminishes instead of being rejected when compared to the attended information. This theory is proposed to explain the sophistication and extensive rigorous processing of the unattended stimuli at different occasions when compared to Broadbent's Selective Filter theory [15]. This detailed approach resulted in adding more layers to explain how the unattended stimuli is operated when compared to the Selective Filter theory claiming that there is an attenuation process instead of a filtered environment for the unattended inputs. This approach gave a rise to claim the possibility of meaningful information extraction from the content in cases of unrelated inputs. In this case, the attenuated channels are analyzed based on their meaning with a recognition of the information that attracts the listener when relevance is recognized by the system that gives a rise to the possibility of an output response production.

1.2.4 Bottom-Up Theory of Perception

The bottom-up perception theory is based on the unidirectional flow of information from the source of information to the relevant locations within the brain through the sensory information communicated from the location. This was initiated by Gibson based on the influence of the environmental stimulus to the brain [20]. In Gibson's approach, optic array picks up consistent information from the environment while interacting with varying environment elements due to sensation that results in perception. The bottom-up approach directly relates to real-time interpretation of the sensed information through the relevant location of the brain. The theory relies on raw data based on direct experience and emphasizes the stimulus itself instead of the pre-existing knowledge or concepts [21, 22, 23]. Information patterns are in the form of

- optic flow with speed and direction,
- texture formation and gradient with the location of the object viewed, and
- visualized objects' function.

An example of such testing is conducted in [24]. The participant data was collected to identify the levels of effectiveness in memory tasks. During each trial, participants were initially instructed about the automatic or self-paced nature of the trial. A stream of images was then viewed by the participants through either a self-paced or automated fashion. Selection of the images by the participants was the next step to demonstrate their recognition memory task capability by using a mouse. The participants' confidence of choice was then analyzed. The data collected from 821 participants indicated that both memory accuracy and memory confidence were greater on self-paced trials compared to the automatic trials. This testing indicated active and self-paced progression to increase the ability of remembering better when images were viewed in comparison to participants having no control over the speed and duration of the presentation of the images [24].

1.2.5 Multistore Model of Memory

Atkinson and Shiffrin introduced the Multistore model of memory in 1968, which is also known as Modal memory. According to this model, there are three storage locations:

1. Sensory register
2. Long-term memory is capable of storing information for an extended period of time with both the timeframe and the storage capacity being indefinite.
3. Short-term memory has the storage capacity for a short period of time (usually up to 30 seconds) with the associated capability of manipulation. There are different systems of memory involved in functioning of the short-term memory with the short period of retention for information pieces.

According to this model, the information flow must be from one storage to another with the retainment at the short-term memory if repetition takes place.

1.2.6 Top-Down Theory of Processing (Indirect, Constructivist)

The central idea behind the Top-Down memory processing approach by George in 1972 was based on the experiences and expectations of an individual's perception to be an active process. The data attained upon perception is correlated with the previously gained knowledge, experiences, expectations, etc. known as the schematic information attained by the brain on the data. The brain uses the existing information to recognize changes based on the new sensory information. Examples of such changes can be dimensions, colors, illusions, distortions, fiction, etc. The three areas of expansion in this area of research included behavioral, computational, and neural aspects [25].

1.2.7 Computational Theory

Marr suggested the Computational theory in 1982 indicating the process of object recognition to be in three consecutive stages that followed a hierarchy [26]:

- *First stage*: Retinal representation as the initial phase of object construction with the two phases being "raw primal sketch" and "full primal sketch". The raw primal sketch is the recognition of the variation in light intensities while the full primal sketch outlines the structure of the shape to derive the 2.5-dimensional representation of the object.
- *Second stage*: The 2.5-dimensional representation of the object is interpreted by the person based on the individual's point of view.
- *Third stage*: The 3-dimensional view of the object is constructed through detailed binocular gap detection and recovery. The gap recovery can be based on the orientation, depth, similarity to other objects, scope, and uniqueness, etc.

This approach was then adopted as an application for artificial intelligence (AI) and image recognition by Marr and Nishihara in 1982 [27]. It is continued to be used in object detection and analysis in AI applications in the research literature [28].

1.3 Processes That Impact Learning

Learning is organized based on the formation of the associated processes by each individual. Organization of the memory structures based on such processing of the information impacts how fast, effectively, and efficiently such information can be utilized based on the access to the memory. This correlates with the storage of the information in the memory based on the approach taken perceived by the stimuli. One such approach is the bottom-up processing approach that is covered in the previous section. The others to be covered in this section include the following:

- Chunking
- Cognitive neuropsychology
- Cognitive neuroscience
- Environment
- Cognitive Psychology
- Gestalt Psychology
- Innate
- Modularity
- Rehearsal
- Schema

- Semantics
- Physiology

The processes that impact learning are not limited to these factors.

1.3.1 Chunking

Chunking is a cognitive term introduced by Miller in 1956 [29]. During the memory processing of active learning of information, the perceived knowledge is broken down into meaningful groups through re-coding them based on their meanings. The grouping of chunks typically relies on the composition of similarity or meaning of the perceived information. These chunks allow the memory to be optimized in several ways [30–31]:

- Decrease the loading by the working memory
- Increase capacity of the working memory
- Increasing learning efficiency
- Extension of the ability to recognize
- Extension of the tasks to be performed later
- Improves long-term memory organization for information in motor sequences, perceived stimuli, and cognitive processing.

Learning strategies that employ chunking include the following:

- **Use of mnemonics.** Structuring memory aids with cues for grouping information. For instance, formation of acronyms and grouping phone number digits examples.
- **Concept formation.** This is the description of how a person learns to form classes during which a higher-order mental process acts on the information that has been perceived through our sensory organs and encoded and stored in the memory [33]. While learning, studying a small set of concepts helps to form a higher-level concept.
- **Rule learning.** Structuring a rule that allows and encourages learning by chunking. An example is following a serial pattern such as alphabet to group the associated letters [32]
- **Abstraction formation.** This is the process of concept formation by recognizing some common set of features and grouping them on the bases of formed concepts on that feature.

A dedicated research interest in this area is how large a chunk can be in such grouping [34]. Noting that the information comes in different types, forms, and dimensions, the goal in this research was to determine the short-term, working memory, and chink sizes. In this observation, the use of several types such as words, colors, numbers, digits, directions, etc. can be helpful in identifying further details of the organization. In this research, the learning process is tested on randomly given words, and it is realized that organizing such randomly given words into meaningful words formed the chunks to be remembered. Hence, for instance, a list of seven words organized into four chunks can be learned quickly and recalled easily.

1.3.2 Cognitive Neuropsychology

Cognitive neuropsychology deals with human cognition that includes neuropsychological methods that include pathology, imaging, developed deficits, simulation, and recording of brain to identify and determine physiological factors playing a role in the learning processes. It is particularly helpful in understanding normal human cognition and representations in the brain. It helps to determine the impaired functioning and normal human cognitive processing based on research on brain damaged patients.

For instance, the research on mathematics learning of children without reading problems is researched as a part of cognitive neuropsychology. It was determined that some children did not experience any challenges with mathematical word-based problems while others demonstrated specific math disabilities without any reading difficulties [35, 36]. Cognitive process's function varied and therefore the mathematical competency is varied within the domain of consideration [37]. For instance, the level of competency varied from calculation to problem solving with the assessment of the specific mathematical process assessed. There are more studies on computation skills than math problem solving skills. The findings in the literature support the approach of mathematical difficulties stated in [38] that supports involvement of multiple cognitive processes in experiencing mathematical difficulties that are reflecting and requiring to understand more cognitive resources.

Psychoneuroendocrinoimmunology is one of the most recent disciplines promoting research on the role of chronic stress to weaken the immune system and therefore causing low-grade inflammation in various parts of the organism [90]. This field focused on the translations of the conceptual framework to clinical medicine that contributes to a more comprehensive approach and emphasizes patient communication for preserving health [91].

The strong role of the immune system in psychosocial and serious disease conditions has been investigated through research with findings of correlation among them. In one relevant study, psychosocial interventions' association with the immune system's functional improvement over time is determined to be significant with a particular observation on the decrease of the proinflammatory cytokines and increase in the immune cell counts [92]. It is suggested that chronic diseases may be treated by using psychosocial interventions as they have shown in research to be effective and relatively affordable for treating such diseases and improving health.

1.3.3 Cognitive Neuroscience

Cognitive neuroscience is the study dealing with the organization of mental structures based on thought and learned information that draws on the psychological, linguistic, philosophy, and computer modeling aspects of research. It consists of techniques used for studying brain functioning. Among the extensive number of tasks with many variational purposes of such tasks of brain, it is likely that none of them is as crucial as the attention. A part of the research that focused on cognitive neuroscience is the research on attention being not a single entity but rather a finite set of brain processes that correlate with each other during actions related to perceptual, cognitive, and motor skill performances [51]. Overall, the role of brain functioning on learning based on neuroscience implications is rather suggestive rather than empirical in establishing a solid connection between the metabolization of oxygen or glucose and the learning of the brain. There are some practical implications that are observed in real-life applications without the lab specific neuroscientific aspects of research [40].

It was Pluto who advised not to force-feed students thousands of years ago; This advice was centered around educators to not force students to learn through information feeding; he instead suggested providing them opportunities for relating learning to interest or evaluating their readiness. Later, Lev Vygotsky introduced the zone of proximal development theory [41, 42]. Vygotsky, a Russian psychologist, introduced a theory that explains children to first develop lower mental functions such as simple perceptions, associative learning, and involuntary attention. These are then improved through social interactions with more advanced peers and adults that eventually lead to the development of higher mental functions such as counting, language, problem solving skills, and memory schemas. The zone of proximal development is the central idea of Vygotsky's theory with the associated cognitive development. In this theory, a child's immediate potential for cognitive growth is mentioned to be bounded on the lower end by self-accomplishments of the child and on the upper end by accomplishments with the help of others, such as a peer or teacher. This region of immediate potential is called the zone of proximal development. In this approach, the child's cognitive skills develop as new tasks are learned to be completed with less and less assistance. Stephen Krashen later included the idea of individualizing and differentiating instructions in the zone of proximal development theory [43] in addition to the negative impact of stress on learning.

The central components of the stress system are in the hypothalamus and the brain stem. They are part of a major neuroendocrine control system called the hypothalamic–pituitary–adrenal axis. One of the two major operational modalities of the stress system is the rapid activation of the sympathetic nervous system through release of adrenaline with the hypothalamic–pituitary–adrenal axis being stimulated in parallel. Corticotropin-releasing hormone and adrenocorticotropic hormone are the two important hormones released from the hypothalamus and the pituitary helping to coordinate the fast behavioral and metabolic response to stress. Many of the behavioral and physical responses occurring during the acute stress response are due to the glucocorticoids that are released by ACTH action on the adrenal gland. The other modality is the key player that handles daily social stress with a long-lasting modulation of social behavior. A variety of circuits in the limbic system take place in this modality, and modulation of behaviors accompanying chronic stress is mainly handled by Corticotropin-releasing hormone [69]. Hence these systems take an active place during the learning process.

The gaming used for learning had shown the impact of individualized instructions that demonstrated the effectiveness of differentiating instructions for the learners with the ability of learners driving the learning results [44]. The benefit of video games is the voluntary involvement of the learners with no frustration or overwhelming conditions. Even though there are no tangible awards necessarily, the motivation of the player with practice over time allows improvement on the content matter and the player experiences the neurochemical feedback to be the pleasure. While it was suggested that dopamine could be the main reason for success in learning, upon review of neurochemical, electrophysiological, and other behavioral evidence in the literature, it was concluded in [45]. that dopamine systems are not needed either to mediate the hedonic pleasure of reinforcers or to mediate predictive associations involved in hedonic reward learning, but they may be rather needed for incentive salience attributions to the neural representations of reward-related stimuli.

Reticular Activating system is a small neural network in the brain that is closely related to the automatic nervous system. The reticular activating system is the first stage of the sensory intake filter applied in learning as a primitive network of cells in the lower brain with the filtered information communicated with the higher brain [45, 46]. This filtration technique is also applied in artificial neural network development in AI applications [47]. Just like it is the case for animals, the Reticular Activating system prioritizes the threads detected for the responses and it generates an environment of fight, flight, or freeze [48]. Hence, the Reticular Activating system has a significant role in learning. Noting that a thread can be in any form of stress causing matters such as punishment or embarrassment of the student, being selected as the last person for succeeding in a role among everyone, and anxiety of work completion in front of everyone, the Reticular Activating system can be a part of any possible scenario in the classroom. The neuroimaging has shown the effectiveness of the Reticular Activating system with the choices of neural signal choices. In the case of triggered curiosity, the sensory input may increase dramatically. Similarly, novelty changes are also shown to impact the neural system triggering with curiosity arising and learning to be improved [40]. There is a dependence of consciousness on the activation of the thalamus by the ascending reticular activating system and on the thalamocortical reentrant circuits [62]. Conscious experience requires rapid bidirectional interactions among distributed neuronal populations.

The low-level subcortical areas such as the hypothalamus, the medial part of the thalamus, or the hippocampus activation of the cortex (namely, the cingulate gyrus) are observed to be triggering emotional experiences according to [64]. Additionally, the cortical area is observed to activate the cingulate cortex upon a stream of thoughts and subsequently the subcortical areas such as the hippocampus and the hypothalamus that result in experiencing emotions. Noting the correlation of stress and the learning experience, the learners' experiences of negative emotions due to increased stress levels impacting these pathways and result in neural network selective reasonings. Clinical examples of patients with lesions in these anatomical regions that display a variety of disturbances in their affective and emotional behavior are given in [64]. Prominent changes in memory are identified in the cases of viral infection affecting the hippocampus and the cerebellum with the associated patients explained to suffer from anxiety and rage episodes [62, 64]. Clinical studies have shown the importance of the role of the anatomic network called Papez circuit introduced in [64]. The Papez initiates in the hippocampus that then continues into the fornix to reach the mammillary body, that through the mammillothalamic tract projects to the anterior nucleus of the thalamus that in turn connects to the cingulum, and after the entorhinal cortex, finally returns to the hippocampus [65]. Clinical observations also showed the importance of the Papez circuit on these brain structures on organizing the memory [62].

In accordance with MacLean's model introduced in [68], Damasio [66] observes a strong tie between the cognition and emotions emphasizing that emotions play a major part in decision-making abilities and behavior and points out due to the tight connections between the limbic system and the prefrontal cortex. The role of emotion-related feedback in behavior and decision making is introduced as a part of the study.

The patients with damage on the prefrontal cortex demonstrate changes in personality with striking emotional and decision-making deficits are well known among the neurologists and neuropsychologists [62]. A small region of the prefrontal cortex called the ventromedial prefrontal cortex is determined to have profound effects on social behavior and social functioning without inducing obvious impairments in cognitive performances [67] based on a study on Phineas Gage, an individual who had a workplace accident in which an iron tamping rod entered and exited his skull.

As a part of the science of cognition, intellectual curiosity is the main driver of motivation in learning. The role of communicating neurons with neural representation of information is a major driver of the brain's functioning. In alignment with this structure, localization of functions and topographic organization of the brain is necessary to observe and understand. One of the major methods used in cognitive science for further understanding underlying physiological reasonings of understanding is the use of fMRI. For instance, understanding learners' ability to solve equations and the associated function localization and topographic organization can be observed by fMRI.

A particular application of cognitive science is understanding the role of brain on concepts' representation of STEM students' abstraction of content knowledge. Comparing the abstract concepts with concrete-object concepts, the abstract concepts cannot be captured easily due to the complexity of the neural network. In such a system, the higher levels of hierarchical cortico-cortical networks require extensive amount of effort for processing abstract concepts [52].

Some of the theories used in cognitive neuroscience such as Hebbian learning, predictive coding, and neuronal recycling:

- *Hebbian learning* integrates the motor, sensory and affective information constitutive of high-level models [53].
- *Predictive coding* model aims to explain visual processing as an integration of the feedback connections made from the higher levels of cortical neural network (such as the visual cortex) to lower-level response levels that are stimulus driven [54].
- *Neuronal recycling* is introduced as a hypothesis to explain individuals recycled cortical areas for processing new cultural inventions that are relevant to the functions the areas evolved to support [55].

fMRI was used for neural mapping of undergraduate and graduate students' conceptions of physics by using [56] that resulted in agreement of the recycling theory results attained in [57] for the same goal. The authors divided physics concepts in [56] into four distinct categories that included algebraic equation representation, casual motion, energy flow, and periodicity. It was discovered that each factor was not only distinct based on the activation processes but also the activation of the primitive processes in the associated cortex regions such as sensorimotor and spatial cognitions were distinct.

Another factor in neuroscience is the plasticity of the nervous tissue that relates to the structure and the function of the tissue in relation to the change in its structure. This is a crucial factor at the level of the neuron or its substructures such as dendrites and dendritic spines along with the levels of neurons such as neuronal networks. Neurotransmitters, neurohormones, and chemical entities such as toxic factors are all playing roles in the plastic reorganization of nervous tissue. This plasticity of the nervous tissue can easily impact learning and psychological processes along with brain structure and function by which emotional experiences can change [70]. The adoption capability of the brain allows nervous systematic changes that can occur due to plasticity and therefore impacts the learning mechanisms. The following are the three aspects of plasticity that can be accounted for understanding the impact of plasticity on learning:

1. *Aging.* Research indicated the nervous tissue, and thus brain, to remain it plasticity up until very high age allowing the ability to learn at high ages [71, 72]. In relation to this factor, an active lifestyle with cognitive activities, social and/or physical activities can help to protect against brain and cognitive deterioration [73–75].
2. *Neural Recovery.* Patient recovery is possible due to the plastic changes in the nervous tissue at both neuronal and neuronal network levels that play a significant role in learning. The malfunctioning neural network due to infarcts, brain trauma, toxic factors, and other influences can be compensated with the gifted healing power of the brain that incorporates the process of plasticity. Hence, in the case of brain damage, either the brain cells themselves recover, or there are other brain cells or aggregates of cells which take over the functions of the damaged part [70, 76, 77]. The functionality of the brain can be compensated for due to the regained ability of the neural network healing. Research has shown the importance of practice as a factor in recovery or compensation of the brain with the indications exist that the brain can be stimulated by dedicated forms of practice/training and thereby recover in the most effective ways [78, 79].
3. *Skill Development.* In relation to cognitive knowledge and the use of motor skills towards normal learning, plasticity is observed to be one of the key factors. For instance, MRI brain images of 12 adults who learned juggling were collected before and after learning juggling in [80]. In comparison to the MRI images of the individuals who did not learn juggling, the learners of juggling determined to have an increase in the grey matter volume located in the bilateral mid temporal area. This was one of the research projects that indicated the role of plasticity in

learning a particular skill. Another study on cab drivers in London, England, has shown the specific brain structuring based on the plastic adoption of the brain due to the environmental conditions and necessities.

Brain development is known to change with maturation, and this clearly impacts cognitive and emotional development. A part of the neuroscientific research is focused on sections such as the medial prefrontal cortex and other regions in the frontal lobes as well as tertiary areas in the parietal lobe that developed during the middle and late adolescence but not fully matured well enough until the third decade of life. Specific areas of the brain are observed to fully develop around birth while other parts become mature during early or late childhood stages as well as early, middle or late adolescence stages. While this development is individual dependent, it may continue probably up to approximately 25 years of age [82, 84]. Girls' brains are observed to mature at a faster pace than those of boys [83]. Each individual's maturation factors can be significantly different from others that depend on psychosocial factors in relation to family, peers with their psychological influences and factors along with school-driven factors that directly play a significant role in the neuroscientific development of the brain due to variable exposures. Experimental results have demonstrated significant changes in such areas of interest [84–87]. Executive functions of the brain relate to executive control, planning, concept shifting, working memory's efficient processing, attentional processing, and the so-called 'self-evaluation' and 'social monitoring' with the development of such functions occurring relatively late around middle and late adolescence periods [88]. In all these developments, the underlying role of genetics in biological development of everyone should not be forgotten. This genetic role expands to the next level of impacting maturation and learning through the protein production of the genes. The structure of the proteins of an individual then guides the development of nervous tissue by serving as cell's structural elements such as receptors and enzymes that relate to the guidance of the cell functioning. Such genetic factors along with environmental factors have roles in learning. Environmental factors help in determining the proper functioning of brain structures and thus higher cognitive, psychological, and social functioning. These changes rely on the brain's elastic structure with its potential to allow changes [70]. There is a need for the brain to be nurtured through social and cognitive stimuli and by psychomotor actions for passing the final maturation [89]. The structures (such as anterior cingulate area) in the prefrontal cortex are thought to have strong connections to the limbic areas, and areas in the hypothalamus and thalamus with their role of enabling an efficient behavioral planning with the motivational and emotional processes. Social monitoring and self-evaluations are thought to be guided by these structures. Hence, the medial prefrontal cortex and its connections to other prefrontal, limbic and subcortical areas of importance in relation to educational learning and the role of motivation and emotion [89].

1.3.4 Environmental Factors

In the case of removal of the threat from the learning environment, the Reticular Activating system acts to have clues about strategies about the focus [48]. There are environmental factors that play a role in such clues that relate to triggering the system for further attention:

- Alteration of voice during a presentation
- Change of font size on materials printed.
- Use of different colors and marking key points on a board or learning materials
- Promote curiosity and focus by attracting students when they enter the classroom.
- Demonstrate the concept in a novel way with actions. For instance, throw a ball in the air to explain gravity.
- Announce upcoming events by using actions that provoke curiosity such as placing a video in the online management system for students to view and be curious about the content.
- Actively participating in students triggering their curiosity towards guessing an upcoming event or lecture.
- Physical learning environment conditions such as school environment, temperature, air quality, and noise.

The following are suggested as the four environmental preferences with the associated cognitive determinants in [39]:

- The *mystery* of entering an environment with an interest in increasing learning and/or interaction.
- The *legibility* that allows ease of becoming aware of usage.
- The *coherence* that allows ease of organizing a setting cognitively utilizing mental actions or processes that acquire knowledge and understanding through thought, experience, and the senses.
- The *complexity* of the environment depends on the capacity of the setting to attract interest and promote activity.

A major inclusion to the learning environmental factors with the advancement of computing and internet technologies is the incorporation of web-based education with different sub-modalities that include synchronous, asynchronous, and hybrid learning. All these different sub-modalities at all different levels of learning starting from kindergarten to graduate level advanced studies as well as professional learning and training have different implications on each learner with different mental development and capacity to the moment that they started their web-based educational experiences.

1.3.5 Cognitive Psychology

Cognitive psychology is a branch of psychology with a focus on exploring the operation of mental processes related to perceiving, attending, thinking, language, and memory, mainly through inferences from behavior [49].

In this area of interest as a reflection on learning, similarities, and differences between behavioral and cognitive conceptions of learning, the cumulative nature of most forms of human learning, active and passive nature of learning, comprehension, the role of prior knowledge, and the role played by cognitive analyses of performance has been explored [50]. In particular, the following are some of the research areas focusing on cognitive psychology and learning:

- Attention
- Categorization
- Decision making
- Human learning
- Information retrieval and processing
- Motor control
- Perception
- Psycholinguistics
- Reading
- Short- and long-term memory with conditions of forgetting

1.3.6 Gestalt Theory

The German word gestalt means shape, form, figure. Different researchers interpreted the application of Gestalt psychology to learning in different ways. It is suggested that this theory can be used in deeper level of understanding or learning that arises from critical thinking of the whole picture. Hence, the involvement of cognitions involves active, rather than passive, processing of information. This theory suggests that learning occurs as a whole rather than a divide-and-concur approach. Therefore, this theory reduced experiences to simple stimulus-response reflections. The stimuli need to be integrated for learning to occur. The associated principles are the following:

- Aiding perception
- Understanding
- Identification
- Memorization
- Recognition

For instance, gestalt theory is applied for understanding creativity. In relation to this application, convergent thinking is commonly defined as linear thinking and is usually used in mathematics and science teacher education programs, while divergent thinking refers to a multitude of solutions and is used mostly in the humanities

[58, 59]. It has been observed that strategic selection in a convergent way after divergent solutions is effective on creativity support [61]. What follows is the application of the learning and application of Gestalt principles like the recommended use of both paradigms for developing creativity [60].

1.4 Organs with Significant Role on Learning with Their Impact on Brain Functioning

The human body is a significantly complicated system with many different factors correlation in addition to the environmental, developmental, social, socio-economical, and genetic factors. The role of multisensory learning theory has shown its significance in memorization in one's own language as well as other languages [103]. The associated use of different sensory receptors includes but not limited to

- sight,
- smell,
- touch, and
- sound.

Among these factors, we consider the primary organs playing a role in learning to be eyes, ears, nose, and hand due to their active participation in their communication with the brain during the learning process that will be covered in this chapter.

1.4.1 Role of Eye on Learning of Brain

As covered previously, the research had shown significant impact of the neural systematic development on plasticity of brain but there appeared to be more investigations needed to determine organs' role on such plasticity that impact learning. The eye is one of frontier organs that significantly impact brain's functioning and learning. Visual development occurs during early-life periods with sensory experience and neural development. The role of eye on plasticity of the brain is observed in literature with impact on learning and brain [93]. In their study, the impact of optical quality with the retinal image limitations on the brain's ability to learn and changes on the eyes' positively impacting learning is observed. In the associated literature, training is the focus of measuring significance of the role of eye on the brain and how it impacts the brain and learning.

Focus on normal adult subjects termed perceptual training is emphasized that is considered as the training modality. Functional factors playing a role on normal adult subjects learning included motion perception [94], contrast sensitivity [95], and object recognition [96]. The quality of the retinal images is limited by the uncorrectable optical aberrations even when sphero-cylindrical lenses are used for alignment of focus [97–101]. Such irregularities termed higher order aberrations can occur throughout the life span [102]. Experimental data in [93] demonstrated eye's optical quality limiting visual improvements from perceptual learning in adults well beyond the critical period as it was explained previously. Hence, as explained previously, a greater degree of adult nervous system plasticity exists even during later stages in life. When the higher order aberrations are corrected, it is shown in [93] that larger visual improvements are not due to the better viewing conditions but rather due to perceptual learning. This conclusion was due to the observations on training in two groups of adults. One group with uncorrected higher order aberrations while the other group had the corresponding corrections. On the group with 50% uncorrected trainees, the contrast sensitivity improvement was observed to be on an average bandwidth of 1.42 octaves more than the perceptual learning level that agreed with the literature results. The group with the corrected higher order aberrations had identical training with the other group and they were determined to have significantly different results from the other group: These group members experienced improvements over their visual performances as well as general benefits for all spatial frequencies produced [93].

1.4.2 Role of the Ear on Learning of Brain

The role of the ear in learning cannot be ignored. Ability to attend and process acoustic signal information in a classroom with the processing of the sound signals has a major impact on learning and the associated achievements [106, 107]. Exposure to high-pitching noise in continuum can cause auditory system challenges and changes in psychological conditions. Sensitivity loss to low-pitching noise due to outer hair cell damage can be possible when a rapid growth of loudness occurs with a reduction on the frequency selectivity. Neurons in the auditory nerve are damaged can be due to the damaged inner hair cells. This damage can cause multiple challenges for the students in the classroom that include but are not limited to the following [104]:

- Weaker pitch perception
- Weaker auditory discrimination
- Impact literacy

Aside from the learner's perspective, external impact of the sound in the classroom is also investigated from an acoustic arrangement of the classroom environment perspective in [105]. In their study, upon the evaluation of the use of a new sound field system with its impact on teaching and learning in elementary school classrooms, the main impact of the improved classroom environment was determined to be the significantly increased levels of performance gained. This was an outcome of both the teacher ratings and student performance on the standardized tests that indicated sound-field systems ability to improve performance of children's understanding of the spoken language. The associated academic attainments did not show any benefits of using such a sound-field system. Hence, classroom acoustics were determined to be a significant factor that influenced the efficacy of sound-field systems with some help for children benefitting from such a system in the classrooms with poorer acoustics for improving listening comprehension. Several environmental noise factors considered in the classroom included the following [108–112]:

- Attention
- Literacy
- Mathematics
- Memory

There are tasks that require high attention levels of cognitive processing such as problem solving, attention, and memory that are determined to be vulnerable to noise exposure [112, 113] that may or may not impact the learning process [105].

In addition to the impact of noise in the classroom that can be considered as internal noise, external noise can be a factor in learning. The impact of such negative acoustic factors in learning has also been tried to be addressed at government level; Acoustical Society of America set standards with the associated acoustical performance criteria, design requirements, and guidelines for schools [114]. It is essential to note that such regulations are not easy to implement, and such efforts require school-based implementations.

The impact of the integration of systems in place can be one of the research areas that can be focused on. For instance, noting that the learner is impacted by the performance of ear on learning, if hearing is negatively impacted by undesirable factors in the learning environment such as other students making noise in the classroom, this would correlate with the nervous system's triggering neuron firing due to the increased personal aggression and limit the learning performance. This increased tension at personal level due to the undesired noise could limit learning with a loss in attention, comprehension, and motivation in collaboration with the other subsystems that include eyes and nose to shift focus to a different path of learning. Such subsystems' distraction from the targeted learning for a short period of time can result in significant loss in the ability to learn in concepts that require continuum of attention such as mathematics and science. It is essential to recognize and incorporate the underlying personal psychological and physiological factors during empirical research when data is collected.

1.4.3 Role of the Hand on Learning of Brain

The role of hands for learning new information with hand-brain coordination and collaboration is fundamentally critical for learning. Regardless of the settings of learning setting that can be anywhere such as a factory setting or a classroom as a part of a lab, using hands in any form, such as typing in a computer or hand-writing information, causes hand-brain interaction with the associated collaboration. This fundamental interaction has been observed in several different ways in research literature.

One area of research exploration on hand-brain coordination is based on neurophysiological research on discrete and continuous hand movements and signals attained from the brain based on the hand movements. Most of the focus on such research is based on either discrete movements or continuous movements in literature (please see for example [116–119]) Changes in the brain activities during motor learning are typically measured by using PET scans. A design tracing activity is typically assigned to the subjects of the studies with the observations on the following [120]:

- Brain areas' identification that involved in performing certain tasks (such as maze and square tracing tasks)
- Investigation of differences and similarities in the brain areas related to dominant and nondominant hand performance
- Examination of the effects in these areas during practicing activities

During the performance of maze and square tracing with the right and left hands, primary motor cortex and anterior cerebellum areas that are related to the motor performances are observed to shift lateralization. The representation of the movement pattern is observed to be abstract instead of being muscle specific just like it is observed in [121] with a possibility of containing sequential, temporal, and/or spatial aspects. During the naive and practiced maze tracing, brain areas are determined to be activated differentially. Unskilled performance involved several right cortical areas and left cerebellum with a shift to the activation to SMA during skilled performance. It is likely that the right cerebellum and left dorsal premotor cortex are associated with the temporal aspects of the task while right dorsal premotor cortex and SMA might be associated with the sequential aspects of the observations during the differential effects in different areas. The involvement of parietal areas was determined to be based on the spatial and attentional aspects while error processing and correction was likely triggered the left cerebellum. It was observed that the processes related to the tested motor learning task activate areas that are likely to code abstract information rather than muscle specific information regardless of the hand used.

The observations that included both the discrete and the continuous hand movements' impact on the neural brain activities with their comparisons were conducted for the first time in [115]. Such a differentiation plays an important role in the characterization of eye-hand coordination in association with the neural brain activities. In such a setting, an example of a discrete movement is reaching to an object while an example of a continuous movement is the movement with tracking action. The experimental design in [115] was based on 28 participants joystick management based on sudden and unexpected patterns presented to them. A rapid transition between the two types of movements around 3 Hertz is identified as a tradeoff between the fixed movements and transitioning from one task to another. The focus of this research was based on the eye-hand coordination while the role of the brain could be incorporated into the role of patterns in the associated eye-hand movements.

1.5 Other Factors

There are many factors that are incorporated into learning of the brain in several ways that include but are not limited to the list below. We leave it to the interested people to research and learn more about such factors as it won't be realistic to cover all such concepts in one chapter.

- Monocular Cues
- Object Recognition
- Pattern Recognition Theories
- Speech/Word Recognition
- Focused/Selective Attention
- Visual Attention
- Divided Attention
- Binocular/Oculomotor Cues
- Perceptual Development Theories
- Perceptual Organization
- Color Perception
- Perception

1.6 Future Research Opportunities and Ideas

The major focuses of many scientific research areas have limitations. The main limitation is based on the comprehensiveness of the observations made on the focus groups instead of the real-life actions during learning. While the most recent research on the work related to brain and learning expanded, the associated models still have weaknesses in the factors incorporated. For instance, if a question is asked in a classroom, the anxiety of the student impacts the ability to answer a question that requires the student to think and respond. What would be the differences in anxiety and motivation levels between pointing at a targeted student to ask a specific question in a classroom versus asking the same question to the entire class without pointing anyone with the associated nervous system triggering and functioning of the brain? What are the neural mechanisms triggered in the brain with the associated successful versus unsuccessful attempts? The performance of an individual could be different in a real-life setting with different anxiety and motivational behaviors, and the research accomplished in such settings is very limited. In this regard, the differentiation between such controlled research and real-life groups needs to be tested based on the actual life conditions. The best future research direction to follow appears to integrate psychological and physiological factors for furthering understanding of learning, memory, and brain functioning.

References

[1] Baars, B. J. and S. Franklin 2003. How conscious experience and working memory interact. Trends Cogn. Sci. 7: 166–172. 10.1016/S1364-6613(03).00056-1.

[2] Cowan, N. 2005. Working memory capacity. Exp. Psychol. 54: 245–246. 10.1027/1618-3169.54.3.245.

[3] Chein, J. M., A. B. Moore and A. R. A. Conway. 2011. NeuroImage domain-general mechanisms of complex working memory span. Neuroimage 54: 550–559. 10.1016/j.neuroimage.2010.07.067.

[4] Ashkenazi, S., M. Rosenberg-lee, A. W. S. Metcalfe, A. G. Swigart and V. Menon. 2013. Neuropsychologia visuo – spatial working memory is an important source of domain-general vulnerability in the development of arithmetic cognition. Neuropsychologia 51: 2305–2317. 10.1016/j.neuropsychologia.2013.06.031.

[5] D'Esposito, M. and B. R. Postle. 2015. The cognitive neuroscience of working memory. Annu. Rev. Psychol. 66: 115–142. 10.1146/annurev-psych-010814-015031.

[6] Kim, C., J. K. Kroger, V. D. Calhoun and V. P. Clark. 2015. The role of the frontopolar cortex in manipulation of integrated information in working memory. Neurosci. Lett. 595: 25–29. 10.1016/j.neulet.2015.03.044.

[7] Chai, W. J., A. I. Abd Hamid and J. M. Abdullah. 2018. Working memory from the psychological and neurosciences perspectives: a review. Front Psychol. 27(9): 401. doi: 10.3389/fpsyg.2018.00401. PMID: 29636715; PMCID: PMC5881171.

[8] Miller, G. A., E. Galanter and K. H. Pribram. 1960. Plans and the Structure of Behavior. New York, NY: Henry Holt and Company; 10.1037/10039-000.

[9] Baddeley, A. and G. Hitch. 1974. Working memory. Psychol. Learn. Motiv. 8: 47–89. 10.1016/j.cub.2009.12.014.

[10] Baddeley, A. 1996. Exploring the central executive. Q. J. Exp. Psychol. A 49: 5–28. 10.1080/713755608.

[11] Baddeley, A. D. 2000. The episodic buffer: a new component of working memory? Trends Cogn. Sci. 4: 417–423. 10.1016/S1364-6613(00)01538-2.

[12] Baddeley, A. and G. Hitch. 1974. Working memory. Psychol. Learn. Motiv. 8: 47–89. 10.1016/j.cub.2009.12.014.

[13] Cowan, N. 2005. Working memory capacity. Exp. Psychol. 54: 245–246. 10.1027/1618-3169.54.3.245.

[14] Lachter, J., K. I. Forster and E. Ruthruff. 2004. Forty-five years after Broadbent. 1958: still no identification without attention. Psychol Rev. 2004 Oct; 111(4): 880–913. doi: 10.1037/0033-295X.111.4.880. PMID: 15482066.

[15] Treisman, A. 1964. Monitoring and storage of irrelevant messages in selective attention. Journal of Verbal Learning and Verbal Behavior 3(6): 449–459.

[16] Costall, A. 2017. 1966 and All That: James Gibson and Bottom-Down Theory. Ecological Psychology 29(3): 221–230. https://doi.org/10.1080/10407413.2017.1330121.

[17] Theeuwes, J. 2010. Top–down and bottom–up control of visual selection. Acta Psychologica 135(2): 77–99.

[18] Gibson, E. J. 1963. Perceptual learning. Annual Review of Psychology 14: 29–56.

[19] Gibson, J. J. 1966. The Senses Considered as Perceptual Systems. Boston: Houghton Mifflin.

[20] Gibson, E. J. 1969. Principles of Perceptual Learning and Development. New York: Appleton-Century-Crofts.

[21] Gibson, J. J. 1979. The Ecological Approach to Visual Perception. Boston: Houghton Mifflin.

[22] Gibson, J. J. and E. J. Gibson. 1955. Perceptual learning: Differentiation or enrichment? Psychological Review 62: 32–41. doi:10.1037/h0048826.

[23] Gibson, E. J. and A. D. Pick. 2000. An Ecological Approach to Perceptual Learning and Development. New York: Oxford University Press.

[24] Kennedy, B. L., S. B. Most, T. Grootswagers and V. K. Bowden. 2024. Memory benefits when actively, rather than passively, viewing images. Attention, Perception, and Psychophysics 86(1): 1–8.

[25] Kveraga, K, A. S. Ghuman and M. Bar. 2007. Top-down predictions in the cognitive brain. Brain Cogn. 2007 Nov; 65(2): 145–68. doi: 10.1016/j.bandc.2007.06.007. PMID: 17923222; PMCID: PMC2099308.

[26] Marr, D. and H. K. Nishihara. 1978. Representation and recognition of the spatial organization of three-dimensional shapes. Proceedings of the Royal Society of London. Series B. Biological Sciences 200(1140): 269–294.

[27] Marr, D. and H. K. Nishihara. 1978. Visual information processing: Artificial intelligence and the sensorium of sight. Technology Review 81(1): 2–23.

[28] Reppa, I. and E. C. Leek. 2019. Surface diagnosticity predicts the high-level representation of regular and irregular object shape in human vision. Atten Percept Psychophys 81: 1589–1608. https://doi.org/10.3758/s13414-019-01698-4.

[29] Miller, G. A. 1956. The magical number seven, plus or minus two: Some limits on our capacity for processing information. Psychological Review 63(2): 81–97.

[30] Simon, H. A. 1974. How big is a chunk? Science 183(4124): 482–488.

[31] Fountain, S. B. and K. E. Doyle. 2012. Learning by chunking. *In*: Seel, N. M. (eds.). Encyclopedia of the Sciences of Learning. Springer, Boston, MA. https://doi.org/10.1007/978-1-4419-1428-6_1042.

[32] Jones, M. R. 1974. Cognitive representations of serial patterns. *In*: Kantowitz, B. (ed.). Human Information Processing: Tutorials in Performance and Cognition. Hillsdale: Erlbaum.

[33] Goetz, C. G. (ed.). 2007. Textbook of Clinical Neurology (Vol. 355). Elsevier Health Sciences.

[34] Simon, H. A. 1974. How big is a chunk? Science 183(4124): 482–488.

[35] Rourke, B. P. and M. A. J. Finlayson. 1978. Neuropsychological significance of variations in patterns of academic performance: Verbal and visual–spatial abilities. Journal of Pediatric Psychology 3: 62–66.

[36] Willcutt, E. G., S. A. Petrill, S. Wu, R. Boada, J. C. DeFries, R. K. Olson et al. 2013. Comorbidity between reading disability and math disability: Concurrent psychopathology, functional impairment, and neuropsychological functioning. Journal of Learning Disabilities 46: 500–516.

[37] Geary, D.C. 2005. Role of cognitive theory in the study of learning disability in mathematics. Journal of Learning Disabilities 38: 305–307.

[38] Fuchs, L. S., D. Fuchs, C. L. Hamlett, W. Lambert, K. Stuebing and J. M. Fletcher. 2008. Problem-solving and computational skill: Are they shared or distinct aspects of mathematical cognition? Journal of Educational Psychology 100: 30–47.

[39] Kaplan, S. and R. Kaplan. 1982. Cognition and Environment: Functioning in an Uncertain World Cognition and Environment: Functioning in an Uncertain World (New York: Praeger, 1982).

[40] Willis, J. 2010. The current impact of neuroscience on teaching and learning. Mind, Brain and Education: Neuroscience Implications for the Classroom, 45–68.

[41] Doolittle, P. E. 1995. Understanding Cooperative Learning through Vygotsky's Zone of Proximal Development.

[42] Vygotsky, L. S. 1978. Zone of proximal development: A new approach. Mind in Society: The Development of Higher Psychological Processes, 84–91.

[43] Krashen, S. D. 1982. Principles and practice. Learning 46(2): 327–69.

[44] Reigeluth, C. M. and E. Schwartz. 1989. An instructional theory for the design of computer-based simulations. J. Comp. Based Instr. 16(1): 1–10.

[45] Samuels, I. 1959. Reticular mechanisms and behavior. Psychological Bulletin 56(1): 1.

[46] Olds, J. and B. Peretz. 1960. A motivational analyses of the reticular activating system. Electroencephalography and Clinical Neurophysiology 12(2): 445–454.

[47] Shim, J. 2005. Automatic knowledge configuration by reticular activating system. pp. 1170–1178. *In*: Advances in Natural Computation: First International Conference, ICNC 2005, Changsha, China, August 27–29, 2005, Proceedings, Part I 1. Springer Berlin Heidelberg.

[48] Raz, A. and J. Buhle. 2006. Typologies of attentional networks. Nature Reviews Neuroscience 7(5): 367–379.

[49] Cognitive Psychology, American Psychology Association dictionary, https://dictionary.apa.org/cognitive-psychology.

[50] Anderson, J. R. 2005. Cognitive Psychology and its Implications. Macmillan.
[51] Pashler, H. E. 1998. The Psychology of Attention (Cambridge, Mass.: The MIT Press, 1998); Raja Parasuraman, ed., The Attentive Brain (Cambridge, Mass: The MIT Press, 1998); and Elizabeth A. Styles, The Psychology of Attention (East Sussex, U.K.: Psychology Press Ltd., 1997).
[52] Friston, K. 2003. Learning and inference in the brain. Neural Networks 16(9): 1325–1352.
[53] Hayes, J. C. and D. J. Kraemer. 2017. Grounded understanding of abstract concepts: The case of STEM learning. Cognitive Research: Principles and Implications 2: 1–15.
[54] Friston, K. and S. Kiebel. 2009. Cortical circuits for perceptual inference. Neural Networks 22(8): 1093–1104.
[55] Hernandez, A. E., H. L. Claussenius-Kalman, J. Ronderos, A. P. Castilla-Earls, L. Sun, S. D. Weiss et al. 2019. Neuroemergentism: A framework for studying cognition and the brain. J. Neurolinguistics. 2019 Feb; 49: 214–223. doi: 10.1016/j.jneuroling.2017.12.010. Epub 2018 Nov 29. PMID: 30636843; PMCID: PMC6326375.
[56] Mason, R. A. and M. A. Just. 2016. Neural representations of physics concepts. Psychological Science 27(6): 904–913.
[57] Dehaene, S. and L. Cohen. 2007. Cultural recycling of cortical maps. Neuron 56(2): 384–398.
[58] Land, M. H. 2013. Full STEAM ahead: The benefits of integrating the arts into STEM. Procedia Computer Science 20: 547–552.
[59] Pehkonen, E. 1997. The state-of-art in mathematical creativity. ZDM 29(3): 63–67.
[60] Smith, K. H. 2024. Gestalt theory and constraints in relation to creativity. *In*: Tillman, D. A. (eds.). Exploring Perspectives on Creativity Theory and Research in Education. Creativity Theory and Action in Education, vol 8. Springer, Cham. https://doi.org/10.1007/978-3-031-55416-2_8.
[61] Putri, R. D. 2019. Group Guidance Based on Gestalt Theory to Develop Creativity in Learning. KONSELI: Jurnal Bimbingan dan Konseling 6(2): 149–156.
[62] Steck, A. and B. Steck. 2015. Brain and Mind: Subjective Experience and Scientific Objectivity. Springer.
[63] Dehaene, S., J. P. Changeux, L. Naccache, J. Sackur and C. Sergent. 2006. Conscious, preconscious, and subliminal processing: a testable taxonomy. Trends in Cognitive Sciences 10(5): 204–211.
[64] Papez, J. W. 1937. A proposed mechanism of emotion. Arch Neurol Psychiatry 38(4): 725–43.
[65] Sullivan, E. V. and A. Pfefferbaum. 2009. Neuroimaging of the Wernicke–Korsakoff syndrome. Alcohol and Alcoholism 44(2): 155–165.
[66] Damasio, A. R. 1996. The somatic marker hypothesis and the possible functions of the prefrontal cortex. Philos Trans R Soc Lond B Biol Sci. 351(1346): 1413–20.
[67] Damasio, H., T. Grabowski, R. Frank, A. M. Galaburda and A. R. Damasio. 1994. The return of Phineas Gage: clues about the brain from the skull of a famous patient. Science 264(5162): 1102–5.
[68] MacLean, P. D. 1949. Psychosomatic disease and the visceral brain; recent developments bearing on the Papez theory of emotion. Psychosom Med. 11(6): 338–53.
[69] Hostetler, C. M. and A. E. Ryabinin. 2013. The CRF system and social behavior: a review. Frontiers in Neuroscience 7: 52243.
[70] Kolb, B. and I. Q. Whishaw. 2008. Fundamentals of Human Neuropsychology (6th ed.). New York: Worth Publishers.
[71] Buonomano, D. V. and M. M. Merzenich. 1998. Cortical plasticity: from synapses to maps. Annual Reviews of Neurosciences 21: 149–186.
[72] Weinberger, N. M. 1995. Dynamic regulation of receptive fields and maps in the adult sensory cortex. Annual Reviews of Neurosciences 18: 129–158.
[73] Bosma, H., M. P. van Boxtel, R. W. Ponds, M. Jelicic, P. Houx, J. Metsemakers et al. 2002. Engaged lifestyle and cognitive function in middle and old-aged, non-demented persons: a reciprocal association?. Zeitschrift für Gerontologie und Geriatrie 35(6): 575–581.
[74] Karp, A., S. Paillard-Borg, H. Wang, M. Silverstein, B. Winblad and L. Fratiglioni. 2006. Mental, physical and social components in leisure activities equally contribute to decrease dementia risk. Dementia and Geriatric Cognitive Disorders 21: 65–73.
[75] Wilson, R. S., C. F. Mendes de Leon, L. L. Barnes, J. A. Schneider, J. L. Bienias, D. A. Evans et al. 2002. Participation in cognitively stimulating activities and risk of incident Alzheimer disease. JAMA 287: 742–748.

[76] Robertson, I. H., T. Manly, J. Andrade, B. T. Baddeley and J. Yiend, 1997. "Oops!": Performance correlates of everyday attentional failures in traumatic brain injured and normal subjects. Neuropsychologica 35: 747–758.
[77] Robertson, I. H. and J. M. J. Murre. 1999. Rehabilitation of brain damage: brain plasticity and principles of guided recovery. Psychological Bulletin 125: 544–575.
[78] Cicerone, K. D., C. Dahlberg, J. F. Malec, D. M. Langenbahn, T. Felicetti, S. Kneipp et al. 2005. Evidence-based cognitive rehabilitation: updated review of the literature from 1998 through 2002. Archives of Physical Medicine and Rehabilitation 86: 1681–1692.
[79] Taub, E. 2004. Harnessing brain plasticity through behavioural techniques to produce new treatments in neurorehabilitation. American Psychologist 59: 692–704.
[80] Draganski, B., C. Gaser, V. Busch, G. Schuierer, U. Bogdahn and A. May. 2004. Changes in grey matter induced by training. Nature 427: 311–312.
[81] Maguire, E. A., D. G. Gadian, I. S. Johnsrude, C. D. Good, J. Ashburner, R. S. J. Frackowiak et al. 2000. Navigation-related structural change in the hippocampi of taxi drivers. PNAS 97: 4398–4403.
[82] Giedd, J. N. 2004. Structural magnetic resonance imaging of the adolescent brain. Annals of the New York Academy of Sciences 77–85.
[83] Giedd, J. N. 2008. The teen brain: insights from neuroimaging. Journal of Adolescent Health 42: 335–343.
[84] Gogtay, N., J. N. Giedd, L. Lusk, K. M. Hayashi, D. Greenstein, A. C. Vaituzis et al. 2004. Dynamic mapping of human cortical development during childhood through early adulthood. Proceedings of the National Academy of Sciences of the United States of America 101: 8174–8179.
[85] Blakemore, S. J. and U. Frith. 2005. The Learning Brain: Lessons for Education. Oxford: Blackwell Publishing.
[86] Paus, L. 2005. Mapping brain maturation and cognitive development during adolescence. Trends in Cognitive Sciences 60–68.
[87] Steinberg, L. 2005. Cognitive and affective development in adolescence. Trends in Cognitive Science 9: 69–74.
[88] Blakemore, S. J. and S. Choudhury. 2006. Development of the adolescent brain: implications for executive function and social cognition. Journal of Child Psychology and Psychiatry 47: 296–312.
[89] de De Jong, T., T. Van Gog, K. Jenks, S. Manlove, J. Van Hell, J. Jolles et al. 2009. Explorations in Learning and the Brain: On the Potential of Cognitive Neuroscience for Educational Science. Springer Science and Business Media.
[90] Aaseth, J., J. Alexander and U. Alehagen. 2021. Coenzyme Q(10) supplementation - In ageing and disease. Mech Ageing Dev. 197: 111521. https://doi.org/10.1016/j.mad.2021.111521. Epub 20210612. PubMed PMID: 34129891.
[91] Furman, D., J. Campisi, E. Verdin, P. Carrera-Bastos, S. Targ, C. Franceschi et al. 2019. Chronic inflammation in the etiology of disease across the life span. Nat. Med. 25(12): 1822–32. https://doi.org/10.1038/s41591-019-0675-0. Epub 20191205. PubMed PMID: 31806905; PubMed Central PMCID: PMC7147972.
[92] Shields, G. S., C. M. Spahr and G. M. Slavich. 2020. Psychosocial interventions and immune system function: a systematic review and meta-analysis of randomized clinical trials. JAMA Psychiat. 77(10): 1031–43. https://doi.org/10.1001/jamapsychiatry.2020.0431. PubMed PMID: 32492090; PubMed Central PMCID: PMC7272116.
[93] Zhou, J., Zhang, Y., Dai, Y., Zhao, H., Liu, R., Hou, F. et al. 2012. The eye limits the brain's learning potential. Scientific Reports 2(1): 364.
[94] Liu, Z. L. 1999. Perceptual learning in motion discrimination that generalizes across motion directions. P Natl Acad Sci. USA 96: 14085–14087.
[95] Sowden, P. T., D. Rose and I. R. L. Davies. 2002. Perceptual learning of luminance contrast detection: specific for spatial frequency and retinal location but not orientation. Vision Research 42: 1249–1258.
[96] Furmanski, C. S. and S. A. Engel. 2000. Perceptual learning in object recognition: object specificity and size invariance. Vision Res 40: 473–484.

[97] Thibos, L. N., R. A. Applegate, J. T. Schwiegerling, R. Webb and V. S. Taskforce. 2000. In Vision Science and its Applications (VSIA) (Santa Fe, New Mexico).

[98] Pagnani, M. 2011. Refractive surgery technology advances may expand options for surgeons, patients. Ocular Surgery News 29(18).

[99] Smirnov, M. S. 1961. Measurement of the wave aberration of the human eye. Biofizika 6:s 776–795.

[100] Howland, B. and H. C. Howland. 1976. Subjective measurement of high-order aberrations of the eye. Science 193: 580–582.

[101] Liang, J. Z., B. Grimm, S. Goelz and J. F. Bille. 1994. Objective measurement of wave aberrations of the human eye with the use of a hartmann-shack wave-front sensor. Journal of the Optical Society of America a-Optics Image Science and Vision 11: 1949–1957.

[102] Brunette, I., J. M. Bueno, M. Parent, H. Hamam and P. Simonet. 2003. Monochromatic aberrations as a function of age, from childhood to advanced age. Invest Ophth Vis. Sci. 44: 5438–5446.

[103] von Kriegstein, K. February 6, 2015. Learning with all these sensors, https://www.mpg.de/8934791/learning-senses-vocabulary. Accessed: March 30, 2026.

[104] Flexer, C. 2004. The impact of classroom acoustics: Listening, learning, and literacy. In Seminars in Hearing (25(2): 131–140). Copyright© 2004 by Thieme Medical Publishers, Inc., 333 Seventh Avenue, New York, NY 10001, USA.

[105] Dockrell, J. E. and B. Shield. 2012. The impact of sound-field systems on learning and attention in elementary school classrooms.

[106] McSporran, E. 1997. Towards better listening and learning in the classroom. Educational Review 49(1): 13–20.

[107] Picard, M. and J. S. Bradley. 2001. Revisiting Speech Interference in Classrooms: Revisando la interferencia en el habla dentro del salón de clases. Audiology 40(5): 221–244.

[108] Cohen, S. 1980. After effects of stress on human performance and social behavior: A review of research and theory. Psychological Bulletin 88: 82–108.

[109] Cohen, S., G. W. Evans, D. S. Krantz and D. Stokols. 1980. Physiological, motivational, and cognitive effects of aircraft noise on children: Moving from the laboratory to the field. American Psychologist 35: 231–243.

[110] Cohen, S., D. S. Krantz, G. W. Evans, D. Stokols and S. Kelly. 1981. Aircraft noise and children: A longitudinal and cross-sectional evidence on adaptation to noise and the effectiveness of noise abatement. Journal of Personality and Social Psychology 40: 331–345.

[111] Cohen, S. and N. Weinstein. 1981. Non-auditory effects of noise on behavior and health. Journal of Social Issues 37: 36–70.

[112] Shield, B. M. and J. E. Dockrell. 2008. The effects of environmental and classroom noise on the academic attainments of primary school children. The Journal of the Acoustical Society of America 123: 133–144. doi:10.1121/1.2812596| issn 0001-4966.

[113] Evans, G. W. and S. J. Lepore. 1993. Nonauditory effects of noise on children: A critical review. Children's Environments 10: 31–51.

[114] Standard, A. N. S. I. 2010. ANSI/ASA S12. 60-2010/Part 1 American National Standard Acoustical Performance Criteria, Design Requirements, and Guidelines for Schools, Part 1: Permanent Schools. USA: Acoustical Society of America.

[115] Coudiere, A. and F. R. Danion. 2024. Eye-hand coordination all the way: from discrete to continuous hand movements. Journal of Neurophysiology.

[116] Henriques, D. Y. P. and J. D. Crawford 2002. Role of eye, head, and shoulder geometry in the planning of accurate arm movements. Journal of Neurophysiology 87(4): 1677–1685.

[117] Hadjidimitrakis, K. 2020. Coupling of head and hand movements during eye-head-hand coordination: there is more to reaching than meets eye. Journal of Neurophysiology 123(5): 1579–1582.

[118] Arora, H. K., V. Bharmauria, X. Yan, S. Sun, H. Wang and J. D. Crawford. 2019. Eye-head-hand coordination during visually guided reaches in head-unrestrained macaques. Journal of Neurophysiology 122(5): 1946–1961.

[119] Jana, S., A. Gopal and A. Murthy. 2017. Evidence of common and separate eye and hand accumulators underlying flexible eye-hand coordination. Journal of Neurophysiology 117(1): 348–364.

[120] Van Mier, H., L. W. Tempel, J. S. Perlmutter, M. E. Raichle and S. E. Petersen. 1998. Changes in brain activity during motor learning measured with PET: effects of hand of performance and practice. Journal of Neurophysiology 80(4): 2177–2199.
[121] Keele, S. W. 1981. Behavioral analysis of movement. Handbook of Physiology, Section 1: The Nervous System (Motor Control) 2: 1391–1414.
[122] Stocco, A., P. Rice, R. Thomson, B. Smith, D. Morrison and C. Lebiere. 2024. An integrated computational framework for the neurobiology of memory based on the ACT-R declarative memory system. Computational Brain and Behavior 7(1): 129–149.
[123] Shiffrin, R. M. and M. Steyvers. 1997. A model for recognition memory: Rem—retrieving effectively from memory. Psychonomic Bulletin and Review 4(2): 145–166.
[124] Blum, K. I. and L. F. Abbott. 1996. A model of spatial map formation in the hippocampus of the rat. Neural Computation 8(1): 85–93.
[125] Weber, M., P. D. Maia and J. N. Kutz. 2017. Estimating memory deterioration rates following neurodegeneration and traumatic brain injuries in a hopfield network model. Frontiers in Neuroscience 11: 623.

Chapter 2

Psychological Factors Driven by Correlation of Brain with Anger

Emre Tokgoz,[1,*] *Sophia Marra*[2] and *Percin Karakol*[3]

2.1 Introduction

Anger is a complex emotional response that incorporates many factors. Darwin's anger definition was driven by the action-reaction understanding of behavior; Anger is an action oriented to counteract an obstacle [52]. The neural and cognitive processes have major roles in anger and aggression. At the neural level, anger can be associated with abnormal functioning of the amygdala and ventromedial prefrontal cortex. At the cognitive level, research outcomes suggested that anger can be associated with excessive attention to anger-related stimuli and impulsiveness [42]. In research, there are certain definitions of anger introduced in the literature [22]:

- *State anger* is defined as an individual's angry feelings experienced at that moment.
- *Trait anger* is an individual's tendency to experience elevations in anger status in various situations.
- Upon experiencing anger, *anger expression* is a measure of how anger is expressed.

[1] School of Engineering and Technology, Farmingdale State College, New York, 11735 USA.
[2] Frank H. Netter School of Medicine, Quinnipiac University, North Haven, CT, 06473, USA.
[3] Başakşehir Cam and Sakura City Hospital Plastic Reconstructive and Aesthetic Surgery, Istanbul, Turkiye.
* Corresponding author: tokgoze@farmingdale.edu

DOI: 10.1201/9781003646044-2

- The tendency to suppress or hold in the angry feelings is called *anger-in.*
- Expressing angry feelings physically or verbally is called *anger-out.*
- Controlling anger by declining to express is called *anger-control.*

The psychology of emotions has been color coded in literature in relation to the wheel of emotions proposed by R. Plutchik. The following is the corresponding high and low intensities based on the basic emotions:

- Anticipation emotion with Interest and Vigilance classifications as low and high intensities.
- Joy emotion with Serenity and Ecstasy classifications as low and high intensities.
- Trust emotion with Acceptance and Admiration classifications as low and high intensities.
- Fear emotion with Apprehension and Terror classifications as low and high intensities.
- Surprise emotion with Distraction and Amazement classifications as low and high intensities.
- Sadness emotion with Pensiveness and Grief classifications as low and high intensities.
- Disgust emotion with Boredom and Loathing classifications as low and high intensities.
- Anger emotion with Annoyance and Rage classifications as low and high intensities.

In relation to emotions, there are many different factors that play a role in interplay between the human brain and psychological development with the associated factors; Some of such factors to be covered in this work include the following [6]:

- The relationship between anger and brain anatomy with a reflection on the psychological state of a person.
- Anger and the associated disorders that impact the psychological states of a person.
- Impact of anger on linguistics with a reflection on the psychological state of a person.
- Impact of anger on the physical well-being of a person.
- Physiological impact of brain during anger and aggression.
- The correlation between anger and age of a person.
- Management of anger with the impact of brain's status during such management.
- Measuring psychological and physiological factors of emotions during activity versus resting states.

Figure 1 below is a representation of Genava Emotions Wheel that is used for measuring emotional reactions to objects, events, and situations. The GEW contains 20 different families of emotions with their intensities [126].

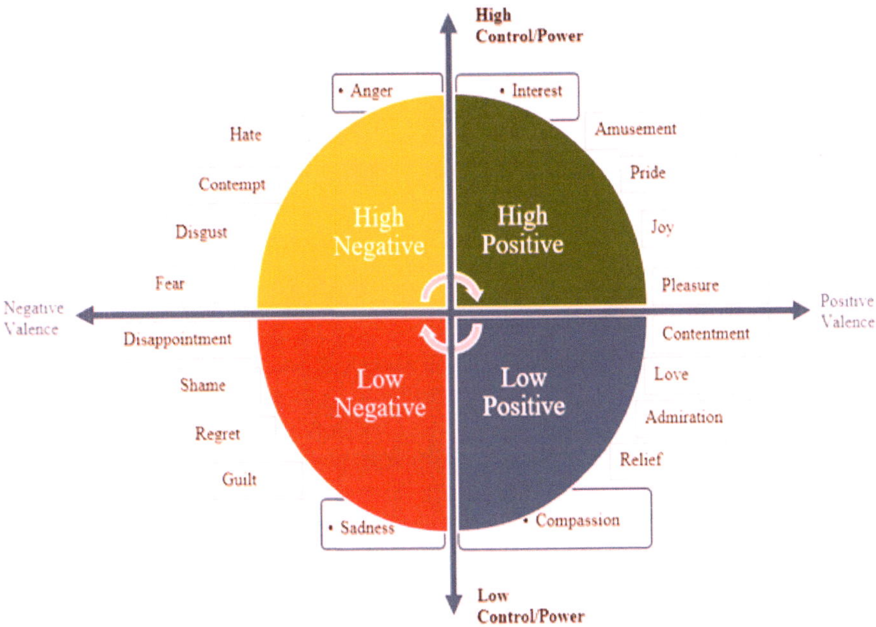

Fig. 1. A representation of Genava Emotions Wheel.

Several theories have been introduced to explain anger [4]:

- Social-Constructionist theory [53]. This theory explains anger as a social syndrome by using a wholistic approach with an assumption that this syndrome cannot be broken down into subclasses of any other factors such as cognitive, physiological, or psychological factors. Anger is situational therefore the definition of a typical anger experience cannot be defined.
- Cognitive-Neoassociationistic theory [54]. An automatic response is generated for reacting to a situation based on a primitive form of anger that depends on a combination of network elements based on memory, thoughts, and feelings along with a strong emphasis on motor-sensor and physiological reactions. The experienced situation evokes personal interpretations, causal attributions, and appraisals that yield to constructing thoughts and feelings that are related to the source of the actual emotional state of anger. What follows is the control and regulation of anger and reactions to it.

Hence, given that anger is a part of emotional behavior, emotion is defined in literature in different ways. The two focus points of emotion consisted of the following:

- *Triggered stimulus conditions*. The major focus of stimulus-triggered approach of emotions definitions focused on the individual's evaluation of the stimulus that causes the emotions to determine the emotion's valence [33].

- *Evoked response conditions*. The individual's subjective evaluation of the feeling determines the valence.

The researchers focused on asymmetric structure of the brain during anger associated conditions during a variety of conditions that consisted of the following:

- Relationship examination between resting EEG and responses to emotion-eliciting stimuli [7]
- Relationship between behavioral and resting EEG activity.
- EEG changes during emotionally evocative situations.
- MRI data collected on brain regional activities for different disorders.
- fMRI data collected on brain regional activities during resting stages for different disorders.

2.2 Impact of Anger on Brain Anatomy with Physiological and Psychological State Effects

The brain acts as the center for peripheral responses and for experiencing emotions. Research on emotions in neuroscientific studies has shown association of certain activities in brain areas that regulate attention, behavioral motivation, and perceived stimulus. The stimulus processing pathway in the brain starts with the stimulus that leads to the response [6]. Anger is agreed to be an emotion that causes approach inclinations by researchers. Anatomical structure of the brain in association with anger has been one of the focus areas of several researchers over the years [1, 5, 7, 11, 17, 18]. The factors that play significant roles in positivity and negativity of affective traits are associated with anger that is driven by personality traits and interaction with others; therefore, the focus of such research is driven by the reaction of each individual to affective trait of the person. Prior research on brain activity on the positive and negative effect was associated with motivation that depended on approach or withdrawal [1]. As a part of one of the research approaches, positive affective traits have been associated with the left frontal cortex while the right frontal cortex has been associated with negative affective traits [27]. An alternative model proposed is that the right region is associated with withdrawal motivation while the left frontal region of the brain is involved in motivation [29, 30]. This approach incorporated uncertainty on the inference of the brain region with the associated motivation approach or positive effect [1]. The method used for testing such traits varied and one of such methods of testing is the use of electroencephalogram (EEG) with the reading of the well-known "alpha power" that helps to measure the rhythms of the brain activities based on the frequency bands of the attained signals [24]. The research results indicated the stability of the human brain's resting frontal asymmetrical activity [28].

The distinguishment between the positive and negative emotions in research were associated with approach- and withdrawal-related motivations respectively [31, 32]. The bias factor in such research is associated with the differentiation in the emotions and the motivation's direction, and anger is an example of such emotion that is noted to be negative in valence that incorporates approach motivation [33, 34]. The researchers considered anger as the underlying testing strategy for both motivations and emotions as a part of the brain's frontal region. In association with all of these research considerations and findings, a natural question of whether the frontal asymmetry reflects the motivational direction and/or the emotional valency. An example to an asymmetrical structure is demonstrated in [35] with the asymmetry relying on the ADHD effects on brain's structural, functional, and common effects. The observations were based on ADHD patients' three cortical regional measurements on the following:

- GMV (an amount of grey matter that lies between the grey-white interface and the pia mater),
- fALFF (fractional amplitude low-frequency fluctuation), and
- ReHo (regional homogeneity).

The authors concluded that the prefrontal cortex may be the primary cause of default mode network and executive network dysfunction in patients with ADHD, noting the prefrontal cortex' connection to emotion, cognition, and executive function. The brain deficits in ADHD determined to be multi-systemic that incorporates fronto-cortical and fronto-subcortical systems that mediate the control of cognition and motivation noting that the laterality of fALFF in ADHD changed in the inferior and middle fronto-temporal and orbitofrontal cortex [35–38].

Furthermore, research on anger is based on several other factors that included but are not limited to the following:

- *Testosterone levels*. Testosterone can modulate levels of various neurotransmitters that show evidence of mediating effects on aggressive behavior based on the recent evidence suggests that these neurotransmitters are involved in processes such as olfaction and arousal and suggestions have been put forward explaining how testosterone may modulate these processes [39].
- Offensive versus Defensive aggression types.
- Unsolvable problem aggression.
- Frustrating events
- Mania and irritability levels
- Individual's disease status
- Enjoyment of anger
- Environmental conditions (such as extreme heat) [40]
- Illnesses
- Anger related disorders

One other area of observation is related to the differences between the resting state of a person and the impact of intrinsic activities on brain regions and emotional state based on the differences generated by such activities. An example of such a study is the identification of the psychological features' impact on emotional picture perception to determine the degree of activity in the resting-state period [43]. The comparison of the resting state periods with the self-related stimuli indicated a significant difference in several task-negative regions that included ventromedial prefrontal cortex, dorsomedial prefrontal cortex, and posterior cingulate cortex (PCC) [44, 45].

Self-relatedness associated specific effects that are distinguished from those of emotions in the subsequent resting-state period involved midline regions, stimulus-evoked effects of self-relatedness during the stimulus period implicated subcortical regions that included periaqueductal gray matter tectum, ventral striatum, amygdala. As a part of the resting state activities, from a psychological perspective, task negative regions are often associated with consciousness [50], episodic memory retrieval [46], self-relatedness [49], and mind wandering [47, 48]. From a physiological perspective, fMRI studies have shown task negative regions being negative blood oxygen level dependent and therefore associated with gamma-aminobutyric acid and neural activity dependent [51].

The correlation between physiology and psychology is investigated on the amygdala part of the brain with emotions evoked in identifying and recognizing threats and relays that generate signals to protect the individual before the cortex based on the intellectuality, reasoning, and judgmental gains [12, 55]. A functional MRI study has shown that human amygdala responses to facial expressions.

2.3 Anger and Disorders

Different levels of anger expressions indicate different levels of disorders. For instance, high levels of trait anger and disturbances in anger expression occur as a part of depression and anxiety. As a part of such observations, left-biased asymmetry of frontal cortical activity is identified to be associated with experiencing anger and expressing outward anger; Hence, different psychopathological disorders can produce various patterns of brain asymmetry [2].

In linkages to such challenges, one can focus on such disorders that are included but not limited to the following to identify anger and impact on brain:

- Anxiety disorders
- Depression
- Eating disorders
- Obsessive-compulsive disorders
- Panic disorders
- Schizophrenia
- Post-traumatic stress disorder
- Neurodevelopmental disorder
- Personality disorders
- Sleep disorders
- Dissociative disorders

- Substance use disorders
- Adjustment disorders
- Bipolar disorders
- Cognitive disorders
- Elimination disorders
- Mood disorders
- Psychosis
- Specific phobia
- Stress
- Delirium
- Difficulty thinking
- Fatigue
- Impulse-control disorder

Anger is identified as a key criterion in five diagnoses within Diagnostic and Statistical Manual of Mental Disorders – 5 (i.e., DSM 5) [9, 57]:

- Bipolar Disorder
- Borderline Personality Disorder
- Disruptive Mood Dysregulation Disorder
- Intermittent Explosive Disorder
- Oppositional Defiant Disorder

Certain cognitive processes relationship to trait anger has been evidenced in relation to

- Interpretation of others' behavior as antagonistic,
- Selectively attending hostile cues, and
- Inclining to ruminate over past incidents that provoked anger

that included the following with particularly high in trait anger [9, 127]:

- Negative rumination
- Distorted reasoning
- Memory biases

The evaluation of psychological states that extend beyond normal behavior and reach psychopathology depends on the following observations [58]:

- *Distress.* Scaling negative feelings at both personal and surrounding levels. For instance, extreme sadness or guilt may be experienced during depression.
- *Deviance.* Deviation from common thoughts, emotions, and behaviors that are expected at the societal level can result in odd conditions. For instance, depression can result in feeling guilty as a deviance reaction.

- *Danger.* A behavior that causes detrimental risk to the individual or society. For instance, self-harming or suicide are the dangers that may be experienced as a part of depression.
- *Dysfunction.* Inability to meet the expected daily functions. For instance, depression may result in people not being willing to leave bed in the morning.

In relation to depression, the brain activity was identified to be at the resting frontal asymmetrical activity with relatively less left than right frontal brain activity of the depressed individuals. Both the self-report indices of depression [59, 60], and individuals identified through clinical interviews [61] indicated the relationship between depression and asymmetrical frontal activity. In addition, individuals with previous clinical depression that had remission status were determined to have a relatively less left frontal activity [62].

Given all the above-listed categories and subcategories with the variational physiological variations in the underlying systems, the brain regions would be expected to react differently with the underlying disease status and challenges. For instance, resting-state functional magnetic resonance imaging is used in [56] for identifying the changes in the functional connectivity strength between amygdala and orbitofrontal cortex to identify such actions' association with trait anger. Left and right amygdala are observed as two separate seed regions to be able to observe sixteen healthy male subjects resting-state conditions. The analysis of the anger measurements correlated to both regions of the amygdala on a voxel-by-voxel basis across all subjects. The authors concluded that anger to be inversely associated with the strength of resting state connectivity between amygdala and contralateral middle orbitofrontal cortex while a stronger association for the right amygdala-left orbitofrontal connection was detected. An opposite pattern of positive correlation with amygdala-orbitofrontal connectivity was detected during the attempt to control anger expressions. Hence, the authors concluded that resting state functional connectivity in a cortico-limbic circuit might subserve stable differences in anger regulation. In relation to this physiological finding, emotional dysregulation disorders such as anxiety, depression, and personality can be furthermore researched in linkage to such findings.

Another area of research interest that relates to psychological well-being is functional neurological disorder (also known as conversion disorder) that is associated with dysregulation of anger-related processes based on patient self-report, observational, and laboratory-based measures in the associated research [19]. In the associated research, preventive actions from adapting, adjusting, or participating in different aspects of life are defined as maladaptive behavior. In relation to this definition, whether maladaptive anger regulation in functional neural disorder can be linked to any specific type of childhood experience is unknown while adverse childhood experiences can be caused from

- genetic factors [68, 69]
- maladaptive anger regulation [63–65],

- suboptimal parental modeling (that may be linked to personality disorders and borderline personality disorder features that can include hostility, resentment, and hitting/shouting [71, 72]) can impact the way of experiencing or expressing negative emotions such as anger [66, 67].

Persons with functional neural disorders are often associated with a history of any of the following during childhood [70]:

- physical abuse
- emotional neglect
- sexual abuse

A personality disorder that is linked to anger is borderline personality disorder (BPD) that is characterized by impulsive behavior, severe mood alterations as well as hardship of developing stable personal relationships. In the associated research, the differences in borderline personality disorder between genders is investigated [23]. The following gender-based factors have been identified:

- Men with BPD are determined to have the following:
  - Impulsive behavior engagement was significantly higher than woman [73]
  - Anger outbreaks [76]
- Women with BPD are determined to have the following:
  - Unstable relationships [75]
  - Affective instability [73–76]
  - Identity disturbance [75, 79]
  - Chronic emptiness feelings [73]
  - No self-harming behaviors [73, 75, 78]
  - Hypersensitivity to external stimuli may result in avoidance and anxiety.
  - A greater tendency to despair is based on higher levels of alexithymia [77].

Overall, the associated research indicated statistically significant results but modest differences among gender in symptom domains and clinical manifestations. The results in relation to aggression suggest that BPD attenuates rather than aggravates the gender difference in aggression usually present in the general population [80]. In one of the studies on BPD, volumetric and cortical thickness measurements were attained from 28 normal and 28 patients with BPD conditions between 18 and 30 years old are attained. The associated MRI and functional MRI image analysis is conducted during both active and resting stages.

fMRI results are also analyzed similarly during the resting state by using the seed-based approach focusing on the amygdala and insula-cortical regions. The following observations are reported [25]:

- No amygdala volume differences are identified between normal and BPD patients.
- Female BPD patients' right amygdala volume was significantly reduced when compared to the healthy group.
- A significant reduction of the cortical thickness was identified for BPD patients.
- Female sub-sample did not have a significant cortical thickness reduction.
- Significant functional connectivity reduction between the amygdala and frontal pole is identified on the BPD patients based on fMRI results.
- Insula functional connectivity was normal for both groups.
- Female BPD patients had significantly higher connectivity between the anterior insula and numerous parts of the brain.

Individuals that experience posttraumatic stress disorder (PSD) have difficulty in managing anger [81, 82]. The severity of the symptoms for this disorder is observed to be correlated with several perspectives:

- Severity of the anger [83],
- Poses a risk of violence towards others [84],
- Crime at legal level [85], and
- Aim to be suicidal within thoughts and behavior [86].

The trauma focused treatments in a variety of studies have been found to be useful as the treatment methods while there are several shortcomings observed along with suggestions as a part of the therapeutic treatment [22]:

- Clinicians tend to focus on other emotions such as shame and guilt rather than anger during the therapy of PSD.
- Post exposure associated treatment also focuses on shame and guilt rather than anger during the therapy.
- Clinicians' training on anger-in and anger-out along with adaptive versus maladaptive anger should be extensive for a comprehensive understanding of the differences.
- Several other aspects of research on anger such as gender and race need to be incorporated into research in order to be able to evaluate the states of the differences and similarities.

MRI neuroimaging of adult patients with PSD demonstrated hippocampal volume reductions. Functional imaging techniques used in research such as PET, SPECT, and fMRI have demonstrated a pattern of a hyperresponsive amygdala operating in the context of attenuated negative feedback from the medial prefrontal cortex and the anterior cingulate gyrus [87]. Noting that the amygdala is involved in mediating negative emotions and physical symptoms associated with autonomic arousal, symptomatology associated with PSD may be explained due to the failure of higher brain regions such as prefrontal cortex for extinguishing conditioned fear responses. Several research results in the literature also indicated marked reductions in hippocampal volume in patients with PSD. The neural network and anger relationship of PSD patients is studied during the resting-state in [16] and fMRI results indicated two prefrontal regions involvement; the left anterior middle frontal gyrus and the right orbitofrontal cortex, with their overall functional connectivity was known to be inversely associated with anger. Furthermore, analysis by using EEG to understand the role of left anterior middle frontal gyrus indicated low-anger patients excited enhanced ability in this area suggesting that left anterior middle frontal gyrus may have a role on protection against anger [16].

2.4 Impact of Anger on Linguistics

Anger-in and anger-out conditions as well as anger-control can all be associated with linguistics and its association with levels of anger. The way patients express their feelings is one way of analyzing their anger-out methods. As a part of the assessments on self-reported outcomes, degree of anger and endorsement of their feelings are asked for expressing their emotions. An example of such a statement is "*I feel like a bomb that is about to explode.*" The operationalization of anger is identified to be dependent on language based on psychometric studies [3]. One of the questioning techniques used widely in research literature to identify behavioral studies is the State-Trait Anger Expression questionnaire [88] that measures seven dimensions of anger.

- *Anger State* and *Anger Trait*
- *Angry Reaction*
- *Anger Control*
- *Anger-in* – inward expressions
- *Anger-out* – outward expression
- *Angry Temperament*
- *Anger prosody* – The means of communication with escalations in sound and tone with angry language as a part of linguistic expressions.

The reflection of linguistics and language used as a part of an individual's daily life can be a factor that may depend on several aspects of one's life [3]:

- Culture [89]
- Social interactions [90–93]
- Environmental circumstances
- Internal processes

Anger can be represented in the language in different forms that can be sometimes difficult to recognize and can vary with the verbalization that also depends on the culture [94]. The nature of adaptive and maladaptive feelings can be reflected in the language in different degrees and ways [3, 95]:

- Mild anger (e.g., words expressing annoyed feeling)
- Moderate anger (e.g., words expressing angry feeling)
- Extreme and intense (e.g., words expressing furious feeling)
- Arousal tended expressions such as "like a bomb."
- The speed of expressing anger in words
- Word senses that relate to anger displays
- Use of words with behavioral outburst
- Provocation words that can be in the form of social provocations or unfair treatment.

Hence, in the variational nature of anger vocabulary the following are observed [9, 96, 97]:

- Words such as furiousness and outrage reflect the phasic bursts of anger.
- Ongoing or tonic anger is reflected based on irritability and irascibility.
- Dispositional rather than situational anger is based on hostility due to frequent pattern occurrence.

The basic dimensions of different configurations can occur in several dimensions [98, 99]:

- Frequency
- Duration
- Intensity
- Mood
- Emotional state
- Temperament

Linguistics in the verbal form also interacts with other forms of actions that can result in aggression. Other forms of categories of aggression in the research literature are introduced as follows [15, 100, 101]:

- *Physical-verbal.* Distinguishes between verbal versus physical ways to express anger.
- *Active-passive.* Active engagement of the aggressor aiming to harm someone while passive aggression remains inactive to cause harm.
- *Direct-indirect.* Direct aggression has face-to-face confrontation between the individuals while indirect is harming by using a secondary source for harming such as another person or object.

2.5 Cardiovascular Activity and Anger

Cardiac activity changes due to anger and aggression-based activities have been observed in the literature [10, 128]. Reduced hearth rate variability and increased heart rate are realized during angry state [129]. Negative valence stimuli triggered elevation in heart rate during aggression in [130] while heart rate variability is lowered during provoking attitude of an opponent [132]. The competitiveness of tasks that female adolescences faced resulted in aggressive behavior towards opponents when compared to a control group during the impact of a competitive task assigned to the groups [131]. As well-known, the interplay between the sympathetic and parasympathetic nervous system branches plays a role in heart rate and the associated rate regulation that indicates a potential role of these systems in the overall anger and aggression regulation [10, 133, 134].

Inductions of anger causing increases in cardiovascular activity are also observed in [135]. Cardiovascular response pattern to challenges faced by an individual has also been found to relate anger [136]. Recognizing stressful events as challenges in the case when the perceived personal resources exceed situational demands observed to cause high cardiac activation coupled with lower vascular resistance while experiencing the same events as threats that are considered as perceived demands greater than resources cause low to moderate cardiac activity coupled with higher vascular resistance [21, 27, 137].

2.6 Age-dependence and Anger

The anger in youth has been an integral part of both clinical and research-based interests [20]. In regards to identification of angry faces, research suggested improved accuracy of detecting angry faces during childhood while angry face detection is further developed during adulthood with a better accuracy of angry face detection for more accurate threat detection [26]. Challenges faced by youth such as aggression, anger, temper loss, defiance, and temper loss are overt externalizing attitudes that take place as a part of youth's mental concerns from both population and clinical setting perspectives [102–105]. These outcomes can be reflected in both school and social life poor outcomes that include but are not limited to the following [106–111]:

- Adverse physical health outcomes
- Anxiety
- Delinquency
- Depression
- Use of substance problems
- School suspension and expulsion
- Severe antisocial behaviors
- Impaired occupational and relational functioning

These issues in return can not only impact the individual in a negative direction but also impact society at large due to worsening over time if early identification and elimination of such issues is underestimated. Regarding the anger-based research on youth, anger is defined to be a temporary valanced emotional state that ranges between mild and intense ranges that cause discomfort with [112, 113]

- Cognitive,
- Behavioral,
- Physiological, and
- Phenomenological

aspects that include but not limited to

- Hostility,
- Arousal, and
- Tendency of aggression.

Aggression can be self-harming or harming others with the angry effect that also relates to irritability that can be hot- or cold-blooded based on variable reactions [113, 114]:

- Proactive
- Reactive
- Impulsive
- Instrumental
- Hostile

Research studies on anger and testosterone of men that are normal adults, adolescents, and convicted criminals during their responses to challenges indicated a correlation between the two phenomena although it is important to note that it is not always easy to associate psychological and physiological conditions due to the complexity of the human nature [14, 115–117]. Salivary testosterone was observed to be significantly correlated with mood of anger and tension and selective attention to angry faces based on the saliva samples collected from both males and females [118]. Evidence in research supports impulsive aggression and/or anger linkage to reduced serotonergic activity with majority of the studies indicating this to be more likely to be a robust case in men than women [119, 120].

Elderly with depressive disorders experienced lower levels of anger when compared to those without depression in a variety of settings [121]:

- Conditions causing hostility.
- Unpredictable disruption.
- Disturbances.
- Embarrassing conditions.
- Personal disrespect.

Depression and anger co-exist for youth as well [123]. The association of anger and depression with suicide risk among adolescent inpatients was also observed

[124], and each have been linked to violent behavior and delinquency [13, 125]. In another research, anger and hostility is observed to extend violence and aggression in domestically violent men [122].

Overall, the existence of anger with depression and other mood disorders has been shown in the literature. Clinically depressed individuals expressed suppression of anger as well as anger directing themselves with potential self-harm. In this case, whether depression is the result of repressed anger does not have an extensive empirical foundation. The connection between anger, depression, and violence requires careful attention as recent mass murder cases by adults and juveniles in the United States, Canada, and Europe have been observed.

2.7 Conclusions and Future Work

Anger is a factor that plays a significant role in the behavior of a person, and the brain is one of the major players in decisions made. Even though the definition of anger focuses on two aspects consisting of triggered stimulus conditions and evoked response emotions, we introduce a broader definition of anger as follows:

> *Anger is an evoked response that occurs upon a combination of stimuli- and emotion-based responses due to an individual's integrated genetic and self-developed physiological conditions as well as the behavioral gained traits that can be due to individual's exposure to self-developed or environmental conditions.*

In this definition, the phrase "environmental factors" does not only include the time-window based moments that the individual is surrounded but also the family and friends. Noting that the brain development is starting to shape during gestation as a highly complex process starting around 12- to 14-week-old embryo, nerve cells are proliferating at the rate of about 15 million per hour with the development of the brain progressing. From about 15 to 20 weeks of pregnancy, the number of brain cells in the cerebral cortex increases rapidly; by the seventh month, the fetus is emitting its own brain waves [41]. Even during this phase, the exposure of the mother's organs to the baby with all positive and negative physiological conditions cannot be underestimated and plays a significant role in the baby's nervous system development at this early stage. The living conditions and external exposure to events and attitudes of others play a role in the gained behavior of the baby starting from childhood. The complex design of human beings with the complex design of the brain significantly impacts the levels of anger.

Some of the psychological factors that can be researched include the following:

- How does the inner voice play a role in anger outcomes?
  - What levels of anger occur with the inner voice?
  - How does it impact anger?
  - What is the significance of past experiences when it comes to impacting the levels of anger and aggression?

- How much does the emphasis on religious belief impact anger and aggression?
  - Can we classify religions' impact on anger and aggression?
  - What religious factors play a role in the decisions made that impact anger and aggression?
- How does memory impact anger?
- What is the interplay between the heart and the brain during anger and aggression? What is the significance of the heart taking control during anger and aggression by using physiological factors and out rule the brain activities?

References

[1] Harmon-Jones, E. 2004. On the relationship of frontal brain activity and anger: Examining the role of attitude toward anger. Cognition and Emotion 18(3): 337–361.

[2] Stewart, J. L., R Levin-Silton, S. M. Sass, W. Heller and G. A. Miller. 2008. Anger style, psychopathology, and regional brain activity. Emotion 8(5): 701.

[3] Alia-Klein, N., G. Gan, G. Gilam, J. Bezek, A. Bruno, T. F. Denson et al. 2020. The feeling of anger: From brain networks to linguistic expressions. Neuroscience and Biobehavioral Reviews 108: 480–497.

[4] Gilam, G. and T. Hendler. 2017. Deconstructing anger in the human brain. Social Behavior from Rodents to Humans: Neural Foundations and Clinical Implications 257–273.

[5] Harmon-Jones, E. and J. Sigelman. 2001. State anger and prefrontal brain activity: evidence that insult-related relative left-prefrontal activation is associated with experienced anger and aggression. Journal of Personality and Social Psychology 80(5): 797.

[6] Statharakos, N., A. J. Alvares, E. Papadopoulou and A. Statharakou. 2022. Psychology of emotions. pp. 21–50. In The Psychology of Anger. Cham: Springer International Publishing.

[7] Harmon-Jones, E. 2004. Contributions from research on anger and cognitive dissonance to understanding the motivational functions of asymmetrical frontal brain activity. Biological Psychology 67(1-2): 51–76.

[8] Harmon-Jones, E., C. K. Peterson and C. Harmon-Jones. 2009. Anger, motivation, and asymmetrical frontal cortical activations. International Handbook of Anger: Constituent and Concomitant Biological, Psychological, and Social Processes 61–78.

[9] Fernandez, E. and S. L. Johnson. 2016. Anger in psychological disorders: Prevalence, presentation, etiology and prognostic implications. Clinical Psychology Review 46: 124–135.

[10] Carver, C. S. and E. Harmon-Jones. 2009. Anger is an approach-related affect: evidence and implications. Psychological Bulletin 135(2): 183.

[11] Sorella, S., V. Vellani, R. Siugzdaite, P. Feraco and A. Grecucci. 2022. Structural and functional brain networks of individual differences in trait anger and anger control: An unsupervised machine learning study. European Journal of Neuroscience 55(2): 510–527.

[12] Yadav, P. K., R. L. Yadav and N. K. Sapkota. 2017. Anger; its impact on human body. Innovare Journal of Health Sciences 4(5): 3–5.

[13] Sigfusdottir, I., G. Farkas and E. Silver. 2004. The role of depressed mood and anger in the relationship between family conflict and delinquent behavior. Journal of Youth and Adolescence 33: 509–522.

[14] Potegal, M., G. Stemmler and C. Spielberger. (eds.). 2010. International Handbook of Anger: Constituent and Concomitant Biological, Psychological, and Social Processes. Springer Science and Business Media.

[15] Ramirez, J. M. and J. M. Andreu. 2006. Aggression, and some related psychological constructs (anger, hostility, and impulsivity) Some comments from a research project. Neuroscience and Biobehavioral Reviews 30(3): 276–291.

[16] Eshel, N., A. Maron-Katz, W. Wu, D. Abu-Amara, C. R. Marmar and A. Etkin. 2021. Neural correlates of anger expression in patients with PTSD. Neuropsychopharmacology 46(9): 1635–1642.

[17] Harmon-Jones, E. 2007. Trait anger predicts relative left frontal cortical activation to anger-inducing stimuli. International Journal of Psychophysiology 66(2): 154–160.
[18] Watson, E. M., J. P. Loveless, A. J. Stephenson, K. L. Bickel, K. A. Lehockey and E. D. Everhart. 2016. The relationship between anger, frontal asymmetry and the BIS/BAS subscales. Journal of Nature and Science 2: e264.
[19] van Dijl, T. L., A. C. Videler, H. P. Aben and W. J. Kop. 2024. Anger regulation in patients with functional neurological disorder: A systematic review. General Hospital Psychiatry.
[20] Evans, S. C., A. R. Karlovich, S. Khurana, A. Edelman, B. Buza, W. Riddle et al. 2024. Evidence base update on the assessment of irritability, anger, and aggression in youth. Journal of Clinical Child and Adolescent Psychology 1–32.
[21] Dienstbier, R. A. 1989. Arousal and physiological toughness: Implications for mental and physical health. Psychological Review 96: 84–100.
[22] Wells, S. Y., T. A. Patel, T. F. Halverson, S. T. LoSavio, L. Morland, T. Wachsman et al. 2024. The impact of trauma-focused psychotherapies on anger: A systematic review and meta-analysis. Psychological Trauma: Theory, Research, Practice, and Policy.
[23] Bozzatello, P., C. Blua, D. Brandellero, L. Baldassarri, C. Brasso, P. Rocca et al. 2024. Gender differences in borderline personality disorder: a narrative review. Frontiers in Psychiatry 15: 1320546.
[24] Watson, D., D. Wiese, J. Vaidya and A. Tellegen. 1999. The two general activation systems of affect: Structural findings, evolutionary considerations, and psychobiological evidence. Journal of Personality and Social Psychology 76(5): 820.
[25] Ceccarini, F., I. Colpizzi and C. Caudek. 2024. Age-dependent changes in the anger superiority effect: Evidence from a visual search task. Psychonomic Bulletin and Review 1–10.
[26] Zaidi, S. S., A. Perveen, M. A. Alam, J. Kishore and U. D. Bhardwaj. 2024. Study to assess the effectiveness of Behavioural Change Communication aid for the anger management in adolescents: A Quasi experimental study in the selected Juvenile aid center, New Delhi. Brain Behavior and Immunity Integrative 100057.
[27] Blascovich, J. and J. Tomaka. 1996. The biopsychosocial model of arousal regulation. *In*: Zanna, M. (ed.). Advances in Experimental Social Psychology 28: 1–51. New York: Academic Press.
[28] Tomarken, A. J., R. J. Davidson, R. E. Wheeler and R. C. Doss. 1992. Individual differences in anterior brain asymmetry and fundamental dimensions of emotion. Journal of Personality and Social Psychology 62(4): 676.
[29] Harmon-Jones, E. and J. J. Allen. 1997. Behavioral activation sensitivity and resting frontal EEG asymmetry: covariation of putative indicators related to risk for mood disorders. Journal of Abnormal Psychology 106(1): 159.
[30] Sutton, S. K. and R. J. Davidson. 1997. Prefrontal brain asymmetry: A biological substrate of the behavioral approach and inhibition systems. Psychological Science 8(3): 204–210.
[31] Lang, P. J., M. M. Bradley and B. N. Cuthbert. 2013. Motivated attention: Affect, activation, and action. pp. 97–135. In Attention and Orienting. Psychology Press.
[32] Watson, D. 2000. Mood and Temperament. Guilford Press.
[33] Lazarus, R. S. 1991. Emotion and Adaptation. Oxford University Press.
[34] Berkowitz, L. 2000. Social representations of anger. Handbook of Cognition and Emotion 411.
[35] He, N., L. Palaniyappan, Z. Linli and S. Guo. 2022. Abnormal hemispheric asymmetry of both brain function and structure in attention deficit/hyperactivity disorder: a meta-analysis of individual participant data. Brain Imaging and Behavior 16(1): 54–68.
[36] Durston, S., J. van Belle and P. de Zeeuw. 2011. Differentiating frontostriatal and fronto-cerebellar circuits in attention-deficit/hyperactivity disorder. Biological Psychiatry 69(12): 1178–1184.
[37] Cubillo, A., R. Halari, A. Smith, E. Taylor and K. Rubia. 2012. A review of fronto-striatal and fronto-cortical brain abnormalities in children and adults with attention deficit hyperactivity disorder (ADHD) and new evidence for dysfunction in adults with ADHD during motivation and attention. Cortex 48(2): 194–215.
[38] Faraone, S. V., P. Asherson, T. Banaschewski, J. Biederman, J. K. Buitelaar, J. A. Ramos-Quiroga et al. 2015. Attention-deficit/hyperactivity disorder. Nature Reviews Disease Primers 15020.
[39] Simpson, K. 2001. The role of testosterone in aggression. McGill Journal of Medicine 6(1).

[40] Blair, R. J. R. 2012. Considering anger from a cognitive neuroscience perspective. Wiley Interdiscip Rev Cogn Sci. 3(1): 65–74.
[41] Ackerman, S. 1992. Discovering the Brain. Washington (DC): National Academies Press (US). 6, The Development and Shaping of the Brain.
[42] Richard, Y., N. Tazi, D. Frydecka, M. S. Hamid and A. A. Moustafa. 2023. A systematic review of neural, cognitive, and clinical studies of anger and aggression. Current Psychology 42(20): 17174–17186.
[43] Schneider, F., Bermpohl, F., Heinzel, A., Rotte, M., Walter, M., Tempelmann, C. et al. 2008. The resting brain and our self: self-relatedness modulates resting state neural activity in cortical midline structures. Neuroscience 157(1): 120–131.
[44] Northoff, G., P. Qin and T. Nakao. 2010. Rest-stimulus interaction in the brain: a review. Trends in Neurosciences 33(6): 277–284.
[45] Busler, J. N., J. A. Yanes, R. T. Bird, M. A. Reid and J. L. Robinson. 2019. Differential functional patterns of the human posterior cingulate cortex during activation and deactivation: a meta-analytic connectivity model. Experimental Brain Research 237: 2367–2385.
[46] Buckner, R. L., J. R. Andrews-Hanna and D. L. Schacter. 2008. The brain's default network: anatomy, function, and relevance to disease. Annals of the New York Academy of Sciences 1124(1): 1–38.
[47] Christoff, K., A. M. Gordon, J. Smallwood, R. Smith and J. W. Schooler. 2009. Experience sampling during fMRI reveals default network and executive system contributions to mind wandering. Proceedings of the National Academy of Sciences 106(21): 8719–8724.
[48] Mason, M. F., M. I. Norton, J. D. Van Horn, D. M. Wegner, S. T. Grafton and C. N. Macrae. 2007. Wandering minds: the default network and stimulus-independent thought. Science 315(5810): 393–395.
[49] Northoff, G., A. Heinzel, M. De Greck, F. Bermpohl, H. Dobrowolny and J. Panksepp. 2006. Self-referential processing in our brain—a meta-analysis of imaging studies on the self. Neuroimage 31(1): 440–457.
[50] He, B. J. and M. E. Raichle. 2009. The fMRI signal, slow cortical potential and consciousness. Trends in Cognitive Sciences 13(7): 302–309.
[51] Northoff, G., M. Walter, R. F. Schulte, J. Beck, U. Dydak, A. Henning et al. 2007. GABA concentrations in the human anterior cingulate cortex predict negative BOLD responses in fMRI. Nature Neuroscience 10(12): 1515–1517.
[52] Tremblay, P. F. and M. J. Wheatley. 2020. Anger. The Wiley Encyclopedia of Personality and Individual Differences: Personality Processes and Individual Differences 31–35.
[53] Averill, J. R. 1983. Studies on anger and aggression: implications for theories of emotion. Am Psychol. 38(11): 1145–1160.
[54] Berkowitz, L. 1990. On the formation and regulation of anger and aggression: a cognitive-neoassociationistic analysis. Am Psychol. 45(4): 494–503.
[55] Novaco, R. W. 2000. Anger, Encyclopedia of Psychology. New York: Oxford University Press.
[56] Fulwiler, C. E., J. A. King and N. Zhang. 2012. Amygdala-orbitofrontal resting-state functional connectivity is associated with trait anger. Neuroreport. 23(10): 606–10. doi: 10.1097/WNR.0b013e3283551cfc. PMID: 22617448; PMCID: PMC4271793.
[57] Fernandez, E. and S. L. Johnson. 2016. Anger in psychological disorders: Prevalence, presentation, etiology and prognostic implications. Clin Psychol Rev. 2016 Jun; 46: 124–35. doi: 10.1016/j.cpr.2016.04.012. Epub 2016 Apr 27. PMID: 27188635.
[58] Davis, T. 2009. Conceptualizing psychiatric disorders using "Four D's" of diagnoses. The Internet Journal of Psychiatry 1(1): 1–5.
[59] Jacobs, G. D. and D. Snyder. 1996. Frontal brain asymmetry predicts affective style in men. Behavioral Neuroscience 110: 3–6.
[60] Schaffer, C. E., R. J. Davidson and C. Saron. 1983. Frontal and parietal electroencephalogram asymmetry in depressed and nondepressed subjects. Biological Psychiatry 18: 753–762.
[61] Allen, J. J., W. G. Iacono, R. A. Depue and P. Arbisi. 1993. Regional EEG asymmetries in bipolar seasonal affective disorder before and after phototherapy. Biological Psychiatry 33: 642–646.

[62] Henriques, J. B. and R. J. Davidson. 1990. Regional brain electrical asymmetries discriminate between previously depressed and healthy control subjects. Journal of Abnormal Psychology 99: 22–31.

[63] Wolff, K. T. and M. T. Baglivio. 2017. Adverse childhood experiences, negative emotionality, and pathways to juvenile recidivism. Crime and Delinquency 63(12): 1495–1521. https://doi.org/10.1177/0011128715627469.

[64] Gardner, F. L., Z. E. Moore and M. Dettore. 2014. The relationship between anger, childhood maltreatment, and emotion regulation difficulties in intimate partner and non-intimate partner violent offenders. Behavior Modification 38(6): 779–800. https://doi.org/10.1177/0145445514539346.

[65] Swopes, R. M., D. V. Simonet, A. E. Jaffe, R. P. Tett and J. L. Davis. 2013. Adverse childhood experiences, posttraumatic stress disorder symptoms, and emotional intelligence in partner aggression. Violence and Victims 28(3): 513–530.

[66] Rhoades, K. A., L. D. Leve, G. T. Harold, J. M. Neiderhiser, D. S. Shaw and D. Reiss. 2011. Longitudinal pathways from marital hostility to child anger during toddlerhood: Genetic susceptibility and indirect effects via harsh parenting. Journal of Family Psychology 25(2): 282–291. https://doi.org/10.1037/a0022886.

[67] Herd, T., A. Brieant, B. King-Casas and J. Kim-Spoon. 2022. Associations between developmental patterns of negative parenting and emotion regulation development across adolescence. Emotion 22(2): 270.

[68] Mick, E., J. McGough, C. K. Deutsch, J. A. Frazier, D. Kennedy and R. J. Goldberg. 2014. Genome-wide association study of proneness to anger. PloS One 9(1): e87257.

[69] Perroud, N., I. Jaussent, S. Guillaume, F. Bellivier, P. Baud, F. Jollant et al. 2010. COMT but not serotonin-related genes modulates the influence of childhood abuse on anger traits. Genes, Brain and Behavior 9(2): 193–202.

[70] Ludwig, L., J. A. Pasman, T. Nicholson, S. Aybek, A. S. David, S. Tuck et al. 2018. Stressful life events and maltreatment in conversion (functional neurological) disorder: systematic review and meta-analysis of case-control studies. The Lancet Psychiatry 5(4): 307–320.

[71] Kaur, M. and M. Sanches. 2023. Parenting role in the development of borderline personality disorder. Psychopathology 56(1-2): 109–116.

[72] Winsper, C., M. Zanarini and D. Wolke. 2012. Prospective study of family adversity and maladaptive parenting in childhood and borderline personality disorder symptoms in a non-clinical population at 11 years. Psychological Medicine 42(11): 2405–2420.

[73] Zanarini, M. C., J. Horwood, D. Wolke, A. Waylen, G. Fitzmaurice and B. F. Grant. 2011. Prevalence of DSM-IV borderline personality disorder in two community samples: 6,330 English 11-year-olds and 34,653 American adults. J. Personal Disord. 25: 607–19. doi: 10.1521/PEDI.2011.25.5.607.

[74] McCormick, B., N. Blum, R. Hansel, J. A. Franklin, D. St John, B. Pfohl et al. 2007. Relationship of sex to symptom severity, psychiatric comorbidity, and health care utilization in 163 subjects with borderline personality disorder. Compr Psychiatry 48: 406–12. doi: 10.1016/j.comppsych.2007.05.005.

[75] De Moor, M. H. M., M. A. Distel, T. J. Trull and D. I. Boomsma. 2009. Assessment of borderline personality features in population samples: is the personality assessment inventory-borderline features scale measurement invariant across sex and age? Psychol Assess 21: 125–30. doi: 10.1037/a0014502.

[76] Tadić, A., S. Wagner, J. Hoch, Ö. Başkaya, R. Von Cube, C. Skaletz et al. 2009. Gender differences in Axis I and Axis II comorbidity in patients with borderline personality disorder. Psychopathology 42: 149. doi: 10.1159/000224149.

[77] Amerio, A., A. Natale, G. B. Gnecco, A. Lechiara, E. Verrina, D. Bianchi et al. 2023. The role of gender in patients with borderline personality disorder: Differences related to hopelessness, alexithymia, coping strategies, and sensory profile. Medicina (Lithuania) 59: 950. doi: 10.3390/medicina59050950.

[78] Marchetto, M. J. 2006. Repetitive skin-cutting: parental bonding, personality and gender. Psychol. Psychother. Theory Res. Pract. 79: 445–59. doi: 10.1348/147608305X69795.

[79] Johnson, D. M., M. T. Shea, S. Yen, C. L. Battle, C. Zlotnick, C. A. Sanislow et al. 2003. Gender differences in borderline personality disorder: findings from the collaborative longitudinal personality disorders study. Compr Psychiatry 44: 90. doi: 10.1016/ S0010-440X(03)00090-7.

[80] Mancke, F., S. C. Herpertz and K. Bertsch. 2015. Aggression in borderline personality disorder: a multidimensional model. Personal Disord. Theory Res. Treat. 6: 278–91. doi: 10.1037/per0000098.
[81] Olatunji, B. O., B. G. Ciesielski and D. F. Tolin. 2010. Fear and loathing: A meta-analytic review of the specificity of anger in PTSD. Behavior Therapy 41(1): 93–105.
[82] Orth, U. and E. Wieland. 2006. Anger, hostility, and posttraumatic stress disorder in trauma-exposed adults: A meta-analysis. Journal of Consulting and Clinical Psychology 74(4): 698–706.
[83] Koenen, K. C., J. M. Stellman, S. D. Stellman and J. F. Jr. Sommer. 2003. Risk factors for course of posttraumatic stress disorder among Vietnam veterans: A 14-year follow-up of American Legionnaires. Journal of Consulting and Clinical Psychology 71(6): 980–986. https://doi.org/10.1037/0022-006X.71.6.980.
[84] Novaco, R. W. and C. M. Chemtob. 2015. Violence associated with combat related posttraumatic stress disorder: The importance of anger. Psychological Trauma: Theory, Research, Practice, and Policy 7(5): 485–492. https://doi.org/10.1037/tra0000067.
[85] Elbogen, E. B., S. C. Johnson, V. M. Newton, K. Straits-Troster, J. J. Vasterling, H. R. Wagner et al. 2012. Criminal justice involvement, trauma, and negative affect in Iraq and Afghanistan war era veterans. Journal of Consulting and Clinical Psychology 80(6): 1097–1102. https://doi.org/10.1037/a0029967.
[86] Dillon, K. H., E. E. Van Voorhees, P. A. Dennis, J. J. Glenn, C. R. Wilks, L. A. Morland et al. 2020. Anger mediates the relationship between posttraumatic stress disorder and suicidal ideation in veterans. Journal of Affective Disorders 269(15): 117–124. https://doi.org/10.1016/j.jad.2020.03.053.
[87] Nutt, D. J. and A. L. Malizia. 2004. Structural and functional brain changes in posttraumatic stress disorder. Journal of Clinical Psychiatry 65: 11–17.
[88] Spielberger, C. D. 1999. Professional manual for the state-trait anger expression inventory-2 (STAXI-2). Odessa, FL: Psychological Assessment Resources.
[89] Immordino-Yang, M. H., X. F. Yang and H. Damasio. 2014. Correlations between social-emotional feelings and anterior insula activity are independent from visceral states but influenced by culture. Frontiers in Human Neuroscience 8: 728.
[90] Damasio, A. and G. Carvalho. 2013. The nature of feelings: evolutionary and neurobiological origins. Nat. Rev. Neurosci. 14: 143–152.
[91] Gilam, G. and T. Hendler. 2017. Deconstructing anger in the human brain. Social Behavior from Rodents to Humans: Neural Foundations and Clinical Implications 257–273.
[92] LeDoux, J. 2012. Rethinking the emotional brain. Neuron 73(4): 653–676.
[93] Panksepp, J. 2010. Affective neuroscience of the emotional BrainMind: evolutionary perspectives and implications for understanding depression. Dialogues in Clinical Neuroscience 12(4): 533–545.
[94] Kircanski, K., M. D. Lieberman and M. G. Craske. 2012. Feelings into words: contributions of language to exposure therapy. Psychological Science 23(10): 1086–1091. https://doi.org/10.1177/0956797612443830.
[95] Buss, A. H. and M. Perry. 1992. The aggression questionnaire. Journal of Personality and Social Psychology 63(3): 452.
[96] Buss, A. H. 1961. Aggression, anger, and hostility. pp. 1–16. *In*: Buss, A. H. (ed.). The Psychology of Aggression. John Wiley & Sons, Inc.. https://doi.org/10.1037/11160-001.
[97] Ramirez, J. M. and J. M. Andreu. 2006. Aggression, and some related psychological constructs (anger, hostility, and impulsivity) Some comments from a research project. Neuroscience and Biobehavioral Reviews 30(3): 276–291.
[98] Fernandez, E., I. Arevalo, R. Vargas and A. Torralba. 2014. Norms for five parameters of anger: How do incarcerated adults differ from the community? International Journal of Forensic Mental Health 13(1): 18–24.
[99] Smith, T. W. 2013. Concepts and methods in the study of anger, hostility, and health. pp. 23–42. In Anger, Hostility, and the Heart. Psychology Press.
[100] Berkowitz, L. 1994. Is something missing? Some observations prompted by the cognitive-neoassociacionist view of anger and emotional aggression. pp. 35–60. *In*: Huesman, R. (ed.), Aggressive Behavior: Current Perspectives. Plenum Press, New York.
[101] Berkowitz, L. 1993. Aggression: Its Causes, Consequences, and Control. McGraw-Hill, New York.

[102] Ghandour, R. M., L. J. Sherman, C. J. Vladutiu, M. M. Ali, S. E. Lynch, R. H. Bitsko et al. 2019. Prevalence and treatment of depression, anxiety, and conduct problems in US children. The Journal of Pediatrics 206: 256–267. https://doi.org/10.1016/j.jpeds.2018.09.021.
[103] Polanczyk, G. V., G. A. Salum, L. S. Sugaya, A. Caye and L. A. Rohde. 2015. Annual research review: A meta-analysis of the worldwide prevalence of mental disorders in children and adolescents. Journal of Child Psychology and Psychiatry 56(3): 345–365. https://doi.org/10.1111/jcpp.12381.
[104] Evans, S. C., K. A. Corteselli, A. Edelman, H. Scott and J. R. Weisz. 2022. Is irritability a top problem in youth mental health care? A multi-informant, multi-method investigation. Child Psychiatry and Human Development 54(4): 1027–1041. Advance online. https://doi.org/10.1007/s10578-021-01301-8.
[105] Freeman, A. J., E. A. Youngstrom, J. K. Youngstrom and R. L. Findling. 2016. Disruptive mood dysregulation disorder in a community mental health clinic: Prevalence, comorbidity and correlates. Journal of Child and Adolescent Psychopharmacology 26(2): 123–130. https:// doi.org/10.1089/cap.2015.0061.
[106] Burke, J. D., R. Loeber, B. B. Lahey and P. J. Rathouz. 2005. Developmental transitions among affective and behavioral disorders in adolescent boys. Journal of Child Psychology and Psychiatry 46(11): 1200–1210. https://doi.org/10.1111/ j.1469-7610.2005.00422.x.
[107] Burke, J. D., R. Rowe and K. Boylan. 2014. Functional outcomes of child and adolescent oppositional defiant disorder symptoms in young adult men. Journal of Child Psychology and Psychiatry 55(3): 264–272. https://doi.org/10.1111/ jcpp.12150.
[108] Loth, A. K., D. A. Drabick, E. Leibenluft and L. A. Hulvershorn. 2014. Do childhood externalizing disorders predict adult depression? A meta-analysis. Journal of Abnormal Child Psychology 42(7): 1103–1113. https://doi.org/10. 1007/s10802-014-9867-8.
[109] Rowe, R., E. J. Costello, A. Angold, W. E. Copeland and B. Maughan. 2010. Developmental pathways in oppositional defiant disorder and conduct disorder. Journal of Abnormal Psychology 119(4): 726–738. https://doi.org/10. 1037/a0020798.
[110] Stringaris, A. and R. Goodman. 2009. Longitudinal outcome of youth oppositionality: Irritable, headstrong, and hurtful behaviors have distinctive predictions. Journal of the American Academy of Child and Adolescent Psychiatry 48(4): 404–412. https://doi.org/10.1097/CHI.0b013e3181984f30.
[111] Odgers, C. L., T. E. Moffitt, J. M. Broadbent, N. Dick Son, R. J. Hancox, H. Harrington et al. 2008. Female and male antisocial trajectories: From childhood origins to adult outcomes. Development and Psychopathology 20(2): 673–716. https://doi.org/10.1017/S0954579408000333.
[112] Eckhardt, C., B. Norlander and J. Deffenbacher. 2004. The assessment of anger and hostility: A critical review. Aggression and Violent Behavior 9(1): 17–43. https://doi. org/10.1016/S1359-1789(02)00116-7.
[113] Sukhodolsky, D. G., S. D. Smith, S. A. McCauley, K. Ibrahim and J. B. Piasecka. 2016. Behavioral interventions for anger, irritability, and aggression in children and adolescents. Journal of Child and Adolescent Psychopharmacology 26(1): 58–64. https://doi.org/10.1089/cap.2015.0120.
[114] Fite, P. J., J. Craig, C. R. Colder, J. E. Lochman and K. C. Wells. 2018. Proactive and reactive aggression. *In*: Levesque, R. J. R. (ed.). Encyclopedia of Adolescence. Springer. https://doi.org/10.1007/978-3-319-33228-4_211.
[115] Archer, J. 2006. Testosterone and human aggression: An evaluation of the challenge hypothesis. Neuroscience and Biobehavioral Reviews 30(3): 319–345.
[116] van Bokhoven, I., S. H. van Goozen, H. van Engeland, B. Schaal, L. Arseneault, J. R. Seguin et al. 2006. Salivary testosterone and aggression, delinquency, and social dominance in a population-based longitudinal study of adolescent males. Hormones and Behavior 50(1): 118–125.
[117] Klinesmith, J., T. Kasser and F. T. McAndrew. 2006. Guns, testosterone, and aggression: An experimental test of a mediational hypothesis. Psychological Science 17(7): 568–571.
[118] van Honk, J., A. Tuiten, R. Verbaten, M. van den Hout, H. Koppeschaar, J. Thijssen et al. 1999. Correlations among salivary testosterone, mood, and selective attention to threat in humans. Hormones and Behavior 36(1): 17–24.
[119] Bond, A. J. and J. Wingrove. 2010. The neurochemistry and psychopharmacology of anger. pp. 79–102. *In*: Potegal, M., G. Stemmler and C. D. Spielberger (eds.). Handbook of Anger. New York: Springer.

[120] Reuter, M. 2010. Population and molecular genetics of anger and aggression: Current state of the art. pp. 27–38. *In*: Potegal, M., G. Stemmler and C. D. Spielberger (eds.). Handbook of Anger. New York: Springer.
[121] Baeg, S., S. K. Wang, I. S. Chee, S. Y. Kim and J. L. Kim. 2011. Anger in elderly patients with depressive disorders. Psychiatry Investig. 8(3): 186–93. doi: 10.4306/pi.2011.8.3.186. Epub 2011 Aug 11. PMID: 21994504; PMCID: PMC3182382.
[122] Maiuro, R. D., T. S. Cahn, P. P. Vitaliano, B. C. Wagner and J. B. Zegree. 1988. Anger, hostility, and depression in domestically violent versus generally assaultive men and nonviolent control subjects. Journal of Consulting and Clinical Psychology 56: 17–23.
[123] Brody, C., D. Haaga, L. Kirk and A. Solomon. 1999. Anger in people who have recovered from depression and never-depressed people. The Journal of Nervous and Mental Disease 187: 400–405.
[124] Cautin, R. L., J. C. Overholser and P. Goetz. 2001. Assessment of mode of anger expression in adolescent psychiatric inpatients. Adolescence 36: 163–170.
[125] Blitstein, J. L., D. M. Murray, L. A. Lytle, A. S. Birnbaum and C. L. Perry. 2005. Predictors of violent behavior in an early adolescent cohort: Similarities and differences across genders. Health Education and Behavior 23: 175–194.
[126] Tikadar, S., S. Kazipeta, C. Ganji and S. Bhattacharya. 2017. A minimalist approach for identifying affective states for mobile interaction design. In Human-Computer Interaction-INTERACT 2017: 16th IFIP TC 13 International Conference, Mumbai, India, September 25–29, 2017, Proceedings, Part I 16: 3–12. Springer International Publishing.
[127] Owen, J. M. 2011. Transdiagnostic cognitive processes in high trait anger. Clinical Psychology Review 31(2): 193–202.
[128] Kruithof, E. S., J. Klaus and D. J. Schutter. 2024. Cerebellar asymmetry of motivational direction: anger-dependent effects of cerebellar transcranial direct current stimulation on aggression in healthy volunteers. The Cerebellum 1–9.
[129] Kreibig, S. D. 2010. Autonomic nervous system activity in emotion: A review. Biol Psychol. 84: 394–421. https://doi.org/10. 1016/j.biopsycho.2010.03.010.
[130] Lorber, M. F. 2004. Psychophysiology of aggression, psychopathy, and conduct problems: A meta-analysis. Psychol Bull. 130(4): 531–52. https://doi.org/10.1037/0033-2909.130.4. 531.
[131] Rinnewitz, L., P. Parzer, J. Koenig, K. Bertsch, R. Brunner, F. Resch et al. 2019. A biobehavioral validation of the Taylor Aggression Paradigm in female adolescents. Sci. Rep. 9: 7036. https:// doi. org/10.1038/s41598-019-43456-4.
[132] Hasan, Y., L. Bègue and B. J. Bushman. 2013. Violent video games stress people out and make them more aggressive. Aggress Behav. 39: 64–70. https://doi.org/10.1002/ab.21454.
[133] Cardinali, D. P. 2017. Autonomic Nervous System: Basic and Clinical Aspects. Cham: Springer International Publishing AG. https://doi.org/10.1007/978-3-319-57571-1.
[134] Levy, M. N. 1971. Brief reviews: Sympathetic-parasympathetic interactions in the heart. Circ Res. 29(5): 437–45.
[135] Herrald, M. M. and J. Tomaka. 2002. Patterns of emotion-specific appraisal, coping, and cardiovascular reactivity during an ongoing emotional episode. Journal of Personality and Social Psychology 83: 434–450.
[136] Suls, J. and C. K. Wan. 1993. The relationship between trait hostility and cardiovascular reactivity: A quantitative review and analysis. Psychophysiology 30: 615–626.
[137] Blascovich, J., W. Mendes, S. B. Hunter, B. Lickel and N. Kowai-Bell. 2001. Perceiver threat in social interactions with stigmatized others. Journal of Personality and Social Psychology 80: 253–267.

Chapter 3

Impact of Brain and Psychology on Social Belonging

Emre Tokgoz,[1,*] *Percin Karakol*[2] and *Sophia Marra*[3]

3.1 Introduction

Social belonging has a fundamentally important role in the development and employment of the brain. Looking in the history of the human being, cooperations among the formed groups of our prehistoric ancestors were the essential drivers of safety and survival in society. This sense of social belonging naturally forms the basis of brain development starting from childhood to extend on the way that life continued to be structured [4]. This sense is described as a natural occurrence of human beings to pervasively find and maintain naturally [24]. The strong correlations between social, physical and mental health cannot be underestimated and require careful attention at individual levels [33]. Natural selections and evolution over time likely impacted skills' development and social norm structures' shaping. In such natural occurrences, a natural question to ask is the impact of environmental changes on social belonging and homesickness; The individuals' need for belonging and homesickness are determined to be directly related while the need for social belonging was also determined to impact homesickness regardless of the number of friends, hence individual's adjustment to the community and attitudes within the community are identified as potential factors in such social belonging [9].

[1] School of Engineering and Technology, Farmingdale State College, New York, 11735 USA.
[2] Başakşehir Cam and Sakura City Hospital Plastic Reconstructive and Aesthetic Surgery, Istanbul, Turkiye.
[3] Frank H. Netter School of Medicine, Quinnipiac University, North Haven, CT, 06473, USA.
* Corresponding author: tokgoze@farmingdale.edu

DOI: 10.1201/9781003646044-3

Human evolution over centuries resulted in different diseases and social changes that impacted social belonging. COVID-19 is one such example of diseases that fundamentally changed social belonging in 2019, and psychosocial strategies adopted by individuals for maintaining resilience in this brand-new social environment were the key factor to scope with challenges such as anxiety, stress, and burnout. Factors such as kindness and compassion in addition to social belonging are determined to be key elements of promoting recovery and growth from the COVID-19 pandemic [10]. The foundations of individuals' social media belonging had been seeded in social belonging up to the beginning of COVID-19 pandemic, and the ideology of belonging to social media networks was likely to be boosted with the beginning of COVID-19.

The two major driving factors of social belonging appear to be internal and external. Internal factors are motivated by the individual's accumulative experiences, and the external factors are others that are independent from the individual's own perceived ideas developed within the social platform. There are several aspects of technology that seem to also be impactful on social belonging in a variety of ways that include but are not limited to the following:

- **Generative Artificial Intelligence.** The use of chatbot frequency is increasing and this is resulting in more human-robot interaction with the involvement of technology [13, 22]. Such dealings naturally impact the human brain development therefore this is a preliminary factor on scaling the changes in brain development and social belonging.
- **Media.** The creative expressions that include but are not limited to videos and images impact the development of the brain and relate to decision making that impact the collective imagination driven by the brain. One approach developed for analysis of such relationships is the predictive coding framework in cognitive science; This approach suggests that the human brain creates and updates the internally developed model of the world to reduce the possibility of surprises and be ready for anticipated future states [27]. The training of the brain through art and media impacts it to incorporate alternative social realities and potentially furthermore expand on the simulated experiences.
- Technological Activities.

In addition to all considerations, social classifications and cultures impact the brain's development and social belongings. For instance, Western culture construction relies on self-independence with the emphasis on uniqueness and separation from others while the Eastern collectivist culture has interdependence and connectivity to others in the society as the signature characteristics. It has been suggested in the literature that socio-economic status impacts the cognitive attitudes; for instance, it has been suggested that higher socio-economic status results in independent and analytically thinking individuals while the lower socio-economic status results in more interdependence of individuals [36]. Therefore, it is possible that the socio-economic status and cognitive tendencies as well as interdependence and connectedness can vary among cultures that also impact the overall brain development at the cultural level.

The rest of this chapter is dedicated to covering the following:

- Impact of Social Exclusion on Brain Development
- Brain and Gender (Age?) Driven Social Belonging Gaps
- Brain Structure and Function for Social Belonging Decision
- Impact of Family Interactions on Brain and Social Belonging
- Impact of Social Bonding on Brain
- Brain Structures of Socially Vulnerable Individuals
- Influence of Social Belonging on Academic Success and Brain
- Diseases Impacting Brain and Social Belonging
- Organizational Improvement and Brain Science
- Injuries Impacting Brain and Socialization
- Brain Activities and Social Pain
- Impact of Stress, Anger, and Emotions on Brain and Social Interactions

3.2 Impact of Social Exclusion on Brain Development

A wide range of factors exist in brain development as a part of social interactions and one of the major areas of such social interactions in early stages of childhood for such progress is school years. Starting with the beginning of the primary school years, a sense of belonging in the social school network and exclusions of individuals in such networks result in not only the well-being of an individual but also the academic performance [1, 39]. In this framework of "need to belong" in which case the individual can be deeply rooted to have the desire to become a part of a social group, it is the peer-interactions and the friendships that children attempt to develop in early stages of such period [37, 40]. The positive and negative impact of individuals needing to belong a social group during this period is believed to have a deeper effect on individuals therefore furthermore research has been conducted on the brain's development during this period on brain changes.

In the research literature, belonging to a group is defined as becoming attached to groups that may relate to the sense of belonging, and belonging has been associated with long-term commitment to relationships for children [6]. Children older than age 5 have been observed to be sensitive to minimal group manipulation; Positive attitudes towards in-group members are determined to be more likely to occur compared to the same attitudes towards out-group members. Older children of ages 8–9 years old who were excluded in online ball games were determined to be negatively impacted causing changes in their moods, sense of control, and self-esteem in laboratory settings [46]. Social exclusions can be strongly impacting adjustment of children to school and therefore highly impactful during the learning process [47].

The positive effect of social inclusion is identified to have a positive impact on cognitive, emotional, and behavioral development while the opposite is observed to lead feelings of loneliness and disliking school [38]. The negative impact of social exclusion on a person in a school can result in detrimental effects on well-being with the stressors actively increasing; Gray matter volume in the whole brain can decrease fast until the age of 15 or 16 with the slow in such decay afterwards [41]. Given the strong association of nervous system's interaction with the brain for such activities, a synaptic reorganization of the brain is identified during such changes, and it is thought (but not proven) to be related to synaptic pruning process [42, 43].

Consequences of social exclusion of children have been studied in the literature with a focus on questioning whether social belonging is a need or not. In this framework, the natural question arose was the consequences of lack of contact with individuals. In the case of infancy, stopping social interactions with the mother can be distressing and result in an increasing negative impact on the children [6]. Prolonged lack of contact is determined to negatively affect health and well-being [45] that appears as a clear indicator of lacking social interactions impacting brain activities negatively. In the case of adults, social exclusion studies are extended to include ignored individuals in online games as well as individuals who ignore making eye contact with each other [48].

Natural occurrence of social exclusion due to unexpected events such as COVID-19 has also been observed to have some effect on youth's education. Such disruption is argued as having a long-lasting effect on some of the people's relationship with the school. Studies reported on emotional-based school absence experiencing people are evaluated in [26] and this absence is reported to impact people who experienced accumulating negative experiences in school such as bullying, peer rejection, etc. in a way that they don't feel like they belong to the school system. Additionally, these people also expressed their lack of control in their school lives.

Cyberbullying is another emerging challenge faced by individuals that relates to brain development and online social exclusion [35]. The typical analysis approach taking place in the research literature is electroencephalography (EEG) measurements for cyber-like bullying analysis. The expansion of internet technologies and advanced smart phones that serve as personal computers with high levels of capabilities result in increasing cyber-bullying challenges. A review of research studies revealed evidence on the relevance of social exclusion and abnormalities with the corresponding links between event related potential and EEG measurements in relation to the following:

- Emotional deviation
- Emotional attention
- Deviance detection
- Response to detection

The major challenge with majority of the relevant literature studies is the small sample sizes with short timeframes of studies that didn't extend to a timeframe to have comprehensive brain functional changes.

3.3 Brain and Gender Driven Social Belonging Gaps

Research literature focused on the foundational understanding of the role of the brain and gender differences in social belonging. Not surprisingly, the male and female brains were driven typically by their own characteristics with the male brain characteristics typically driven by masculine features while the female brain characteristics were characterized by feminine features while some overlap has been observed between the two genders [2, 55]. Placing this basic knowledge in the center of social belonging, brain characteristics are the main drivers of social belonging choices made by individuals, and these characteristics have been measured by neuroscientific brain researchers.

A recent cognitive neuroscience study (in 2019) identified the difficulty of distinctly categorizing brain of the genders into groups by looking at cognitive and temperamental characteristics [56]. A comprehensive study in 2021 with key findings indicated hardly any structural differences between genders' brains. In terms of task completion and functioning, another research concluded that there are no major differences in verbal, emotional, and spatial processing while reproducible brain activation of women and men based on the functional Magnetic Resonance Image results [2].

Psychological factors such as the following can also be direct or indirect contributors to social belonging decisions that need to be further investigated through research with their impact on the brain functioning:

- Personal relationships
- Social skills
- Self-esteem
- Perceptions of others
- Cognitive biases
- Workplace situations
- Societal pressure

Social identity development occurs at an individual level throughout the lifetime, and a part of this development relates to gender. Rejection sensitivity that relates to measuring intensity of reactions to rejection is one approach followed by researchers [62]. The level of sensitivity is determined to be more common in minority groups and associated with negative early experiences along with chronic stereotyping exposure and social identity threats [21]. One area of such social belonging of rejection sensitivity can be measured can be seen in some workplaces; Competitive and historically male dominated institutions/organizations may have high levels of rejection sensitivity that are marked with gender imbalances and relates to reduced self-confidence and withdrawal from potential support mechanisms in such places [63, 64]. Expectancy-based value approach researchers belongingness to be a key factor in occupational choices and engagement; A part of such research focused on organizational psychology infrastructures that causes existence and persistence of gender gaps in organizations that can be also linked to occurrences of underrepresented groups in workplaces [2].

3.4 Brain Structure and Function for Social Belonging Decisions

Adolescence is the maturation period for brain's function and structural progression with a gradual decrease in the cortical grey matter volume and progressive white matter density and volume increase [14, 49]. Such changes along with the on-going physiological changes are likely to relate myelination of long-range axonal pathways, intra-cortical connections, and neural system's synaptic pruning [50, 51]. There are certain systems that play primary roles in outcomes:

1. The affective system assigning emotional and social significance to perceptual stimuli for supporting approach and avoidance responses that includes the following [52]:
   - Amygdala
   - Striatum
   - Nucleus
   - Accumbens
   - Hypothalamus
   - Ventrolateral prefrontal cortex
   - Ventral anterior cingulate
   - Anterior insula
2. The cognitive-regulatory system consisting of the following areas [53]:
   - orbitofrontal
   - dorsomedial
   - ventral pre-frontal cortex

   Social stimuli actively trigger changes in this network to evaluate meaning and context to regulate emotional reactions and inhibition of impulsive attitude [53].
3. The last region is centered around a core network of regions with their impact on mental mediation with the assignment of mental states consisting of the following [54]:

- Medial pre-frontal cortex
- Temporal junction
- Superior temporal sulcus

The role of brain to adopt individuals to be social beings is apparent based on its abilities to regulate human behavior. Negative social experiences that can be recognized as social pain and their relationship to cortical responses are studied in the literature [57, 58]. The bilateral medial prefrontal cortex of the brain as well as anterior cingulate cortex and bilateral insula are found to be associated with social evaluation and self-identity [59, 60]. Regarding other activities of these regions, they are also linked to social pain or physical pain [18, 61]. In this setting, another role of the anterior cingulate cortex is to take place in not only "error evaluation" but also possibly taking place in the Go/No-Go actions. This integrated design of the brain contains partial explanations on social belonging and associated decisions.

The structure of the brain has been studied in the United Kingdom based on the databank compromised of approximately 40 thousand individuals' neuroimaging and behavioral data [3]. Breaking down the data set into three social groups consisting of sports teams, religious groups, and social clubs, structural analysis indicated variations in the following areas that are tightly linked to social belonging as the substrates change in these areas:

- Ventromedial prefrontal cortex
- Fusiform gyrus
- Anterior cingulate cortex

Functional connectivity analysis among the groups indicated a difference between the social club group and the other two groups with an emphasis on the importance of limbic network and default mode network [3]. The changes in the default mode network have also been reported by other researchers in relation to social cognition as a part of their studies that relate to prefrontal cortex and anterior cingulate cortex [64]. In support of the other research literature, default mode network appears to be a neural signature of social belonging [3, 65].

3.5 Impact of Family Interactions on Brain and Social Belonging

Initiated from the infant stage, interactions of a baby with family members have an important impact on the social development and personal development of the infant. Particularly, back-and-forth conversations between family members and the infant that can be helpful in brain's language skill regional development [66]. This is likely to help the baby to initiate early development skills towards social interactions and learn how to progress such skills with others in the future from the babyhood years.

As a part of family interactions, the impact of parenting on child and adolescent social-emotional development with the changes in brain activities attracted researchers. The notion of positive parenting that is associated with high levels of warmth, sensitivity, and involvement can be directly linked to a child's cognitive, social, and emotional behavior [68]. On the contrary, negative parenting is associated with harsh attitudes, criticism, and high negative effects of the parents during parenting. In a more specific manner, there are certain brain-based functional associations of parents' impact on child and adolescent brain development that is likely to impact social behavior. The corresponding neural network activations determined to be impacted include the following [67]:

- Cognitive control
- Reward processing
- Emotional arousal

- Emotion regulation
- Social-Emotional information processing

Some of the key findings that relate to neural network activation in relation to positive/negative parenting included the following:

- In the case when parenting incorporated more negativeness than positiveness towards youth, emotional neural networks turned out to have higher activation in response to negative parenting.
- Reward processing networks are identified to have higher activation during lower positive parenting during monetary rewarding.
- Cognitive control networks during cognitive control tasks are predicted to have lower activation due to lower positive parenting and there was less adaptivity in neural responses to parent-specific stimuli.
- Some of the studies that determined the above-mentioned results to be associated by psychopathology and gender.

The functional connectivity between parenting and brain functionality is researched based on gender differences. This inner connectivity is studied in relation to amygdala and other regions during negative emotional stimuli management. The connectivity levels between Amygdala with emotion processing and emotion regulation regions are impacted negatively during negative parenting [69]. Observations on regular human development suggest Amygdala and medial prefrontal cortex connectivity to be more negative as maturity progresses, and higher negative parenting on youth over time indicated a similar negativity in this connectivity suggesting that negative parenthood on a child and early development of a child into adulthood may have similar effects on a child. In relation and agreement to such results, those children who experienced social stress also appeared to have more "mature" behavior than typical behavior [70]. In one of the studies, Amygdala and parietal lobule are determined to have more positive connection with each other for a 10-year-old boy in comparison to girls as a result of negative parenting therefore furthermore investigation is needed for understanding differences between genders when it comes to the social impact of family on brain activities of a child [71].

3.6 Impact of Social Bonding on Brain

Social bonding differences can impact the role of grain that may depend on groups, ages, environments, activities, and relationship status to name a few. Among these several studies dived into more detail on the impact of social bonding from both psychological and physiological perspectives, and we are going to give several examples of different findings in this section.

One area of focus in social bonding research is the mere belonging in which case small socially connected cues are formed between two persons or a group in a domain with a specific task to be performed. In such cases, experiments indicated a mere sense of social connectedness is determined to enhance motivation towards achievement. In such experiments, individuals are provided with group tasks for completion. Similar studies provide evidence supporting the mere belonging ideology with adoption of the persons to goals and motivation of others as their own motivational goals. Belonging to an intellectual community can be driven by the sense of belonging based on the motivation of individuals [7]. In this regard, enhancing social connectedness in meaningful ways may boost motivation and achievements in both school and work settings [20].

One of the physiological ingredients impacting social bonding is the oxytocin neurohormone with extensive research investigations conducted in this area [29]. The human body is a complex system with hormones signaling molecules with the activation of messengers for regulating the organism's behavior and physiology. Oxytocin's role as a means of social bonding has been investigated in a variety of ways as a part of the research literature:

- Modulation of social cognition
- Influencing parental behaviors
- Facilitation of bonding between different species
- Impact on biological buffers that link to isolation

Similarly, vasopressin is another neurohormone regulating social bonding with its impact on several interactive social bonding activities that include the following. Produced in the hypothalamus and released by the pituitary gland, vasopressin hormone not only regulates the water balance and blood pressure but also actively takes place in social behavior and bonding in monogamous mammals. Stimulation of vasopressin can occur through visual activities that relates to communication of the visualization system's interaction with the brain. It can be released in ways that include physical contact, seeing a loved one, and shared experiences.

- Long term attachment and bonding between partners
- Feelings of trust and social bonding
- Regulation of body's response to stress with the promotion of resilience and emotional stability. Under stressful conditions, vasopressin may buffer the negative effects of stress on relationships and strengthen emotional conditions.

Vasopressin is also impactful on long term social bonding as a part of nurturing love through visual stimuli and physical contact. Some of the activities that can be effective in releasing vasopressin to deepen emotional bonds and fortifying long term connections include the following:

- Physical affection that can include kissing, hugging, and cuddling that promotes vasopressin release.
- Time spent together promoting connection and intimacy between individuals.
- Open and honest communication between individuals can be impactful.
- Alignment of shared goals and values between individuals.

Vasopressin, often hailed as the "bonding hormone," holds the key to fortifying long-term connections and nurturing commitment in love. By comprehending how visual stimuli and physical contact influence vasopressin release, couples can deepen their emotional bonds and cultivate lasting relationships. Investing time in quality moments, embracing physical affection, fostering open communication, and aligning shared goals and values contribute to the intricate dance of love and connection.

Research on facial recognition, memory, empathy, intergroup identification, and pro-social behavior using event-related potential (ERP) and f-MRI reporting studies with a focus on neural mechanism activities indicated the major role of neural responses in a variety of combinations in different manners. ERP studies revealed variational levels of neural activities based on behavior and perception while f-MRI studies also indicated similar results with the dynamically interacting brain regions that do not operate individually during perception and behavior [32].

3.7 Brain Structures of Socially Vulnerable Individuals

The social status of an individual can impact that person's brain structure. An individual can be identified as socially vulnerable that can lead to neurocognitive disorders if one of the following may be experienced by the person:

- Chronic social stress
- Low socioeconomic status
- Poor social adoption

The neural and cognitive research on socially vulnerable populations is limited. One of the studies explored the impact of neural power and connectivity signatures on characterization of the executive performance of healthy but socially vulnerable individuals [8]. This empirical study on 76 participants (with equally distributed control group and socially vulnerable participants) aimed to identify the impact of chronic stress and social disadvantage on cognition by measuring executive functioning and EEG. The following findings are attained:

- Delta-phase synchrony to be lower in control group than the socially vulnerable group.
- Socially vulnerable groups had alike characteristics such as years of schooling and delta-phase synchrony that are identified as the best predictors of belonging to the group.
- Changes in slow-wave neural connectivity and executive functioning are determined to occur individuals that are exposed to chronic stress due to lack of education and socio-economic factors.

This area of research appears to be requiring more investment in furthermore development.

3.8 Influence of Social Belonging on Academic Success and Brain

Academic success has been associated with the correlation of feelings with belonging academic performance, motivation, and dropout rates. For instance, in one of the studies, belonging and academic performance relationships determined the connection between academic achievement in college and level of school belonging in high school [11]. It may be the case that learners can be sensitive to information diagnostic of the quality of their social connections. Another research study investigated how minority and white students' performances are affected by the knowledge of success levels of different ethnic groups in their chosen fields [73]. In this study researchers implemented experimentation to test how the impact of belonging undermines the motivation and achievement of learners that are negatively characterized in academic settings. In the corresponding experimentation, students were led to believe that they may have few friends in an intellectual domain with such knowledge to have no impact on white students while black students' potential and sense of belonging was reduced. This finding is found to be consistent with another research that determined influential impact of the role models (that includes those that are covered in the literature and teachers) on students to look up to with their impact on students and how they define what is attainable [74].

Research also focused on the competencies and the sense of belonging from a skill set and ability perspectives [75]. A set of skills can help an individual to relate with others and develop a sense of identity while competencies can play roles in expected group consistencies, cultural values, and social norms. Social competency maturation can also be central to emotional and social learning approaches that can impact academic achievements such as collaboration [76].

3.9 Diseases Impacting Brain and Social Belonging

There are psychiatric and neurological conditions that play a role in social belonging and brain. Stroke, schizophrenia, depression and anxiety, dementia such as Alzheimer's disease, and neurodegenerative diseases such as Parkinson's disease can dramatically impact the social activities [77]. There is extensive research on such diseases' impact on the brain pathways and functioning.

Parkinson's disease is one of the top diseases that comes to mind that makes a connection between brain impacting diseases and social belonging. It is a result of dopaminergic neuron degeneration in the basal ganglia that can lead to the following [12, 16]:

- Motor symptoms such as tremor, rigidity, bradykinesia, and postural instability
- Non-motor symptoms such as insomnia, fatigue, and cognitive impairment
- Autonomic dysfunction such as orthostatic hypotension

In a qualitative study, authors investigated social adoption process of individuals with Parkinsons disease and categorized the process of social belonging process of results into four:

- **Comprehension.** Understanding the consequences of the diseases in daily life with taken medication, their side effects, and prognosis is the core portion of the disease.
- **Acceptance.** This is the part with the person's accepting the consequences of going through the disease and underlying conditions with the continuous changes' occurrences.

- **Adaption.** Changing the role from ordinary to the changing dynamic conditions of the disease is the process of the individual's self-adaptation to the ever-changing conditions.
- **Balancing.** The norms of life change as the disease brings in new conditions that the individual needs to balance after comprehending the changes to adopt the dynamically changing conditions.

The level of social belonging is border lined by comprehending, accepting, adopting, and balancing the corresponding actions required by the disease.

Dementia such as Alzheimer's disease impacts brain pathways that include memory, language, and reasoning that includes the following [15]:

- **Rate of neuron functioning.** The rate of neurons' stopping to function properly, losing connectivity with other neurons, and eventual death occur much faster than usual.
- **Early stages.** Neurons between entorhinal cortex and hippocampus are damaged in which case hippocampus shrinks.
- **Plaques and Neurofibrillary tangles.** There are plaques that disrupt cell functioning, and neurofibrillary tangles occur due to abnormal accumulations of a protein.
- **Integrated system dysfunction.** The brain pathways dysfunction due to improper functioning of the interconnected neurons.

There are many other diseases that impact the brain with its consequential impact of social belonging and the neural system has an integral role in these diseases that cannot be underestimated.

3.10 Organizational Improvement and Brain Science

Expectation States Theory is an area of research interest that links to social belonging and organizational behavior based on individual's social hierarchical positions that relate to social status carrying emotions. In this theory, the investigations focus on the hypothesis of higher statutes producing positive emotions while lower statutes leading to negative emotions [78]. Hence, the role of social hierarchy as a part of organizational improvement is investigated in relation to brain functioning.

Establishment and processing of social hierarchy appeared to change as identified during neuroimaging observations that included but not limited to the following:

- Ventral striatum [79]
- Prefrontal cortex [80]
- Anterior insula [81]

Parts of the brain that are determined to be activated the most during comparison of hierarchical belonging comparisons included ventral striatum, occipital/parietal cortex, and parahippocampal cortex that suggest attentional resource recruitment while perceiving higher social hierarchical status [79]. There are studies in support of decision-making processes differentially being affected by emotional evaluations from lower and higher social hierarchical positions. In [25], it has been observed that evaluative information given by positions of higher hierarchy compared to lower hierarchy impact decisions made that forms a linkage between social-emotional behavior and social status that appears to agree with the Expectation States Theory, and furthermore investigations are needed in different settings. Negative evaluations from higher positions caused higher activation in brain regions that included right precentral gyrus, right IFG, left fusiform, left SMA, and bilateral IOG. Positive evaluations received from lower positions when compared to higher positions also caused high activations in these regions except for left fusiform [25].

3.11 Injuries Impacting Brain and Socialization

A group brain pathology that is associated with the term Acquired Brain Injury occur after birth that includes the following [17, 28]:

- Stroke
- Tumors
- Infections
- Intoxication
- Hypoxia
- Traumatic brain injuries
- Poisoning

Social interaction manifestation of traumatic brain injuries can be particularly impactful in social interactions in multiple ways:

- **Social cognition.** Comprehension of social rules, norms, and others' emotions can be challenging due to traumatic brain injuries.
- **Emotional regulations.** Controlling emotions can be challenging and can result in inappropriate responses to others.
- **Communication.** Verbal and comprehensive communication can be impaired that can be linked to difficulties in understanding language and information processing.
- **Controlling impulses.** Frontal lobe damage can result in impulse control challenges that make it difficult to though and action filtering.
- **Focus and attention.** Participation in and following conversations in social environments can be difficult due to lack of concentration.
- **Memory.** Recalling prior social interactions and conversational content can be challenging due to memory deficits.
- **Social Isolation.** Behavioral and social challenges can lead to embarrassment and uncomforting mindset for the individual to show interest in social environments that can result in social isolation.

- **Behavioral changes.** The levels of irritability, uncooperativeness, and disengagement can be boosted due to traumatic brain injuries.
- **Quality of life reduction.** Social dysfunction impacts relationships with friends, family, relatives, and society overall.
- **Academic challenges.** Academic performance can be impaired.
- **Occupational challenges.** The ability to execute tasks, performing expected job tasks and occupational functioning can be challenged.

3.12 Brain Activities and Social Pain

A wide range of social pain can be experienced during daily activities that include but are not limited to ostracization by classmates, rejection during an interview, isolation feeling, negative prototyping, bullying, loneliness, social exclusion, and rejection that are typically consequences of emotional distress experienced losses or disruptions during social events [19]. Social pain can be followed by a social loss or rejection in which case the existence of an overlap between neural circuity that takes place in physical pain and social pain [82].

In one of the empirical studies, a group of individuals who recently experienced an undesired relationship breakup were given their ex-partner's pictures to think back about their rejection experience. In another task they received painful heat stimulation. These two tests resulted in observing the anterior insula, anterior cingulate cortex, and sensory related regions (S2 and posterior insula) experiencing increased activities demonstrating common neural region activations during both the rejection and physical pain conditions [83].

The impact of rejection-acceptance process was measured during an empirical study with the participants tested on negative social evaluation based on the received rejection feedback from evaluators [84]. In the study, the participants were made to believe that the evaluator was listening to their interview and providing feedback that included pre-selected responses to indicate rejection, neutral, and acceptance every 10 seconds. The participants rated their feelings after receiving each feedback that led to the results indicating that negative feedback receiving participants felt significantly worse than the other participants with the activity increase observed in bilateral anterior insula and anterior cingulate cortex regions. In [85], participants appeared to also have increased activities in the bilateral anterior insula and anterior cingulate cortex regions when images with the themes of acceptance and rejection were shown to them. Another study on measurement of participants reactions to videos displaying disapproving facial expressions elevated activities in the anterior cingulate cortex region of the participants [86]; this impacted the participants even though the disapproval was not targeting them.

Another area of social pain investigations was conducted to understand the neuroscientific impact of social loss on individuals. The loss of first-degree relatives is examined by providing pictures of a deceased first-degree relative and the pictures of a stranger to measure the neural activity differences between them. The results agreed on the activation of the bilateral anterior insula and anterior cingulate cortex regions one more time [87]. Loss of an unborn child is identified to cause higher levels of anterior cingulate cortex activation compared to those mothers who gave birth to a healthy child when the pictures of smiling babies are shown to them.

Additionally, factors that increase physical pain, such as inflammatory activity, are identified to increase experiences of social pain as well. Noting that inflammatory activity is the body's first line of defense against illness and infection that is known to increase physical pain sensitivity, it is identified to increase feelings of social disconnection and neural sensitivity to social pain in an empirical study [82]. Determining whether decreasing physical pain also decreases social pain is also investigated in the same study. For instance, social support that targets reducing social pain and loneliness is questioned to reduce physical pain as well. The empirical results indicated lower pain ratings in the social support conditions during either one of the sessions that consisted of hand holding with their partners or viewing their pictures [82].

3.13 Impact of Stress, Anger, and Emotions on Brain and Social Interactions

There are well known and studied mechanisms that trigger brain activities due to experienced stress, anger, and emotions that directly relate to social interactions. The following are some of the well-known brain activity related factors:

- **Stress.** Chronic stress is impactful in functioning and structure of the brain that impacts memory, decision making, and emotional regulations. Such stress triggers cortisol that reduces hippocampus' functionality that causes memory-based challenges as well as an increased activity in amygdala activities resulting in emotional reactions that are impactful in social interactions. Pre-frontal cortex is a major contributor to executive functioning such as decision making and impulse control. Given the integrated nature of the body, regulation of stress through a top-down approach with amygdala and hypothalamus is critical for its success. There are excitatory amino acids and glucocorticoids playing key roles in regulation processes, along with mediators such as endocannabinoids and brain-derived neurotrophic factors. The dynamic nature of stress regulation results in a continually changing pattern of gene expression mediated by epigenetic mechanisms involving histone modifications and CpG methylation and hydroxymethylation in addition to the retrotransposon activities that may cause alterations in genomic stability. Structural changes in plasticity and vulnerability of the brain provide a basis for understanding the nature of anxiety and depressive disorders as well as age-related cognitive decline [34].
- **Anger.** In the complex network of brain, amygdala and prefrontal cortex are the key players of the anger related activities. Amygdala acts as a threat detector triggering fight or flight response while prefrontal cortex regulates the emotions and suppresses impulsive reactions. The neural systems involved in mediating the basic response to threat, amygdala-hypothalamus-periaqueductal gray, are regulated by several regions of frontal cortex; orbital, medial and ventrolateral

frontal cortex [40]. An increase in reactive aggression is determined to be associated with increasing gradated response within medial frontal cortex occurrence [31].

- **Emotional regulation.** The regions that play active roles in the regulation of emotions include orbital, medial and ventrolateral frontal cortex [88]. There are frontal cortex regions that are identified to impact interneurons within the amygdala that leads to reducing amygdala activities [89]. It may be the case that emotional suppression may cause such activities due to the frontal amygdala suppression [90]. Given the interconnectedness of emotions, anger, and stress, it is not so surprising to have interconnectedness of neuroscience and brain activities.

References

[1] Raufelder, D., N. Neumann, M. Domin, R. C. Lorenz, T. Gleich, S. Golde et al. 2021. Do belonging and social exclusion at school affect structural brain development during adolescence?. Child Development 92(6): 2213–2223.

[2] Rippon, G. 2023. Mind the gender gap: The social neuroscience of belonging. Frontiers in Human Neuroscience 17: 1094830.

[3] Kieckhaefer, C., L. Schilbach and D. Bzdok, 2023. Social belonging: brain structure and function is linked to membership in sports teams, religious groups, and social clubs. Cerebral Cortex 33(8): 4405–4420.

[4] Allen, K. A. 2019. Making sense of belonging. InPsych 41(3): 8–13.

[5] Ali, S., R. P. Rohner, P. A. Britner and A. Jahn, 2024. Longing for belonging: Feeling loved (or not) and why it matters. Family Relations.

[6] Over, H. 2016. The origins of belonging: Social motivation in infants and young children. Philosophical Transactions of the Royal Society B: Biological Sciences 371(1686): 20150072.

[7] Walton, G. M., G. L. Cohen, D. Cwir and S. J. Spencer. 2012. Mere belonging: the power of social connections. Journal of Personality and Social Psychology 102(3): 513.

[8] Lanfranco, R. C., F. dos Santos Sousa, P. M. Wessel, Á. Rivera-Rei, T. A. Bekinschtein, B. Lucero et al. 2024. Slow-wave brain connectivity predicts executive functioning and group belonging in socially vulnerable individuals. Cortex 174: 201–214.

[9] Watt, S. E. and A. J. Badger. 2009. Effects of social belonging on homesickness: An application of the belongingness hypothesis. Personality and Social Psychology Bulletin 35(4): 516–530.

[10] Slavich, G. M., L. G. Roos and J. Zaki. 2022. Social belonging, compassion, and kindness: Key ingredients for fostering resilience, recovery, and growth from the COVID-19 pandemic. Anxiety, Stress, and Coping 35(1): 1–8.

[11] Osterman, K. F. 2000. Students' need for belonging in the school community. Review of Educational Research 70(3): 323–367.

[12] Rosengren, L., A. Forsberg, C. Brogårdh and J. Lexell. 2021. Social belonging as the main concern for achieving life satisfaction when adapting to parkinson's disease. International Journal of Environmental Research and Public Health 18(16): 8653.

[13] Arnold, M. L. 2022. From Nervous System Changes to Systemic Change in Science: Investigating Sex Differences in the Impact of Maternal Immune Activation on Offspring Brain Development and Behavior, and Building Social Justice in Science Through Pedagogy and Community Initiatives to Increase Belonging (Doctoral dissertation, University of California, Berkeley).

[14] Lamblin, M., C. Murawski, S. Whittle and A. Fornito. 2017. Social connectedness, mental health and the adolescent brain. Neuroscience and Biobehavioral Reviews 80: 57–68.

[15] Johnson, J. C., C. R. Marshall, R. S. Weil, D. E. Bamiou, C. J. Hardy and J. D. Warren. 2021. Hearing and dementia: from ears to brain. Brain 144(2): 391–401.

[16] Schapira, A. H., K. R. Chaudhuri and P. Jenner. 2017. Non-motor features of Parkinson disease. Nature Reviews Neuroscience 18(7): 435–450.

[17] Mahar, C. and K. Fraser. 2011. Barriers to successful community reintegration following acquired brain injury (ABI). International Journal of Disability Management 6(1): 49–67.

[18] Somerville, L. H., T. F. Heatherton and W. M. Kelley. 2006. Anterior cingulate cortex responds differentially to expectancy violation and social rejection. Nature Neuroscience 9(8): 1007–1008.

[19] Jiao, Z., J. Song, X. Yang, Y. Chen and G. Han. 2024. Social pain sharing boosts interpersonal brain synchronization in female cooperation. Acta Psychologica 243: 104138.

[20] Walton, G. M. and G. L. Cohen. 2011. A brief social-belonging intervention improves academic and health outcomes of minority students. Science 331(6023): 1447–1451.

[21] Aronson, J. and M. S. McGlone. 2009. Stereotype and social identity threat. Handbook of Prejudice, Stereotyping, and Discrimination 153–178.

[22] Pani, B., J. Crawford and K. A. Allen. 2024. Can generative artificial intelligence foster belongingness, social support, and reduce loneliness? A conceptual analysis. Applications of Generative AI 261–276.

[23] Brance, K., V. Chatzimpyros and R. P. Bentall. 2024. Social identity, mental health and the experience of migration. British Journal of Social Psychology.

[24] Baumeister, R. F. and M. R. Leary. 2017. The need to belong: Desire for interpersonal attachments as a fundamental human motivation. Interpersonal Development 57–89.

[25] Guo, Z., Z. Xing, L. Liu, J. W. Schwieter and H. Liu. 2024. An fMRI study on how decisions are influenced by affective evaluations from different social hierarchical positions. Cognitive Neurodynamics 1–16.

[26] Hamilton, L. G. 2024. Emotionally based school avoidance in the aftermath of the COVID-19 pandemic: Neurodiversity, agency and belonging in school. Education Sciences 14(2): 156.

[27] Bortolotti, A., A. Conti, A. Romagnoli and P. L. Sacco. 2024. Imagination vs. routines: Festive time, weekly time, and the predictive brain. Frontiers in Human Neuroscience 18: 1357354.

[28] Irwin, L. G. and D. G. Fortune. 2014. Schools-based interventions for reducing stigmatization of acquired brain injury: the role of interpersonal contact and visible impairment. Archives of Clinical Neuropsychology 29(2): 194–205.

[29] Jacob, S. 2024. Oxytocin and its links through scientific lineage. Comprehensive Psychoneuroendocrinology 100230.

[30] Röhr, A. K., N. Kohn, R. Bergs, B. Clemens, A. Lampert, M. Spehr et al. 2024. Increased anger and stress and heightened connectivity between IFG and vmPFC in victims during social interaction. Scientific Reports 14(1): 8471.

[31] Lotze, M., R. Vei, S. Anders and N. Birbaumer. 2007. Evidence for a different role of the ventral and dorsal medial prefrontal cortex for social reactive aggression: An interactive fMRI study. Neuroimage 34(1): 470–478. doi: 10.1016/j.neuroimage.2006.09.028.

[32] Peng, K., L. Tan and X. Zhang. 2024. The neural mechanism underlying differentiated in-group versus out-group face recognition and memory, identification, empathy and pro-social behavior: evidence from fMRI and ERP studies. Journal of Integrative Neuroscience 23(3): 48.

[33] Doyle, D. M. and B. G. Link. 2024. On social health: history, conceptualization, and population patterning. Health Psychology Review 1–30.

[34] McEwen, B. S., N. P. Bowles, J. D. Gray, M. N. Hill, R. G. Hunter, I. N. Karatsoreos et al. 2015. Mechanisms of stress in the brain. Nature Neuroscience 18(10): 1353–1363.

[35] Mills, L., C. Driver, L. T. McLoughlin, T. E. Anijärv, J. Mitchell, J. Lagopoulos et al. 2024. A systematic review and meta-analysis of electrophysiological studies of online social exclusion: evidence for the neurobiological impacts of cyberbullying. Adolescent Research Review 9(1): 135–163.

[36] Kühnen, U. and S. Kitayama. 2024. Culture, social class and the dynamics of the self. KZfSS Kölner Zeitschrift für Soziologie und Sozialpsychologie 1–18.

[37] Neel, C. G. O. and A. Fuligni. 2013. A longitudinal study of school belonging and academic motivation across high school. Child Development 84(2): 678–692.

[38]Ramm, G., M., Prenzel, J., Baumert, W., Blum, R., Lehmann, D., Leutner,... and U. Schiefele. 2006. *PISA 2003: dokumentation der erhebungsinstrumente*. Waxmann.

[39] Baumeister, R. F. and M. R. Leary. 2017. The need to belong: Desire for interpersonal attachments as a fundamental human motivation. Interpersonal Development 57–89.
[40] Baumeister, R. F. 2012. Need-to-belong theory. Handbook of Theories of Social Psychology 2: 121–140.
[41] Brain Development Cooperative Group. 2012. Total and regional brain volumes in a population-based normative sample from 4 to 18 years: the NIH MRI Study of Normal Brain Development. Cerebral Cortex 22(1): 1–12.
[42] Huttenlocher, P. R. 1994. Synaptogenesis in human cerebral cortex.
[43] Gabriel, E. T., R. Oberger, M. Schmoeger, M. Deckert, S. Vockh, E. Auff et al. 2021. Cognitive and affective Theory of Mind in adolescence: developmental aspects and associated neuropsychological variables. Psychological Research 85: 533–553.
[44] Dunham, Y., A. S. Baron and S. Carey. 2011. Consequences of "minimal" group affiliations in children. Child Development 82(3): 793–811.
[45] House, J. S., K. Landis and D. Umberson. 1988. Social relationships and health. Science 241: 540–545. (doi:10.1126/science.3399889).
[46] Abrams, D., M. Weick, D. Thomas, H. Colbe and K. M. Franklin. 2011. On-line ostracism affects children differently from adolescents and adults. British Journal of Developmental Psychology 29(1): 110–123.
[47] Wesselmann, E. D., F. D. Cardoso, S. Slater and K. D. Williams. 2012. To be looked at as though air: Civil attention matters. Psychological Science 23(2): 166–168.
[48] Cullerton-Sen, C. and N. R. Crick. 2005. Understanding the effects of physical and relational victimization: The utility of multiple perspectives in predicting social-emotional adjustment. School Psychology Review 34(2): 147–160.
[49] Giedd, J. N. and J. L. Rapoport. 2010. Structural MRI of pediatric brain development: what have we learned and where are we going? Neuron 67: 728–734.
[50] Lovatt, D., U. Sonnewald, H. S. Waagepetersen, A. Schousboe, W. He, J. H. C. Lin et al. 2007. The transcriptome and metabolic gene signature of protoplasmic astrocytes in the adult murine cortex. Journal of Neuroscience 27(45): 12255–12266.
[51] Blakemore, S. J. 2012. Imaging brain development: the adolescent brain. Neuroimage 61(2): 397–406.
[52] Guyer, A. E., J. S. Silk and E. E. Nelson. 2016. The neurobiology of the emotional adolescent: From the inside out. Neuroscience and Biobehavioral Reviews 70: 74–85.
[53] Nelson, E. E., E. Leibenluft, E. B. McClure and D. S. Pine. 2005. The social re-orientation of adolescence: a neuroscience perspective on the process and its relation to psychopathology. Psychological Medicine 35(2): 163–174.
[54] Schurz, M., J. Radua, M. Aichhorn, F. Richlan and J. Perner. 2014. Fractionating theory of mind: a meta-analysis of functional brain imaging studies. Neuroscience and Biobehavioral Reviews 42: 9–34.
[55] Rippon, G. 2019. The gendered brain: The new neuroscience that shatters the myth of the female brain. Random House 3–24.
[56] Hyde, J. S., R. S. Bigler, D. Joel, C. C. Tate and S. M. van Anders. 2019. The future of sex and gender in psychology: Five challenges to the gender binary. American Psychologist 74(2): 171.
[57] Eisenberger, N. I. and M. D. Lieberman. 2004. Why rejection hurts: a common neural alarm system for physical and social pain. Trends in Cognitive Sciences 8(7): 294–300.
[58] Tomova, L., K. Tye and R. Saxe. 2021. The neuroscience of unmet social needs. Social Neuroscience 16(3): 221–231.
[59] Amodio, D. M. and C. D. Frith. 2006. Meeting of minds: the medial frontal cortex and social cognition. Nature Reviews Neuroscience 7(4): 268–277.
[60] Lieberman, M. D., M. A. Straccia, M. L. Meyer, M. Du and K. M. Tan. 2019. Social, self (situational), and affective processes in medial prefrontal cortex (MPFC): Causal, multivariate, and reverse inference evidence. Neuroscience and Biobehavioral Reviews 99: 311–328.
[61] Lieberman, M. D. and N. I. Eisenberger. 2015. The dorsal anterior cingulate cortex is selective for pain: Results from large-scale reverse inference. Proceedings of the National Academy of Sciences 112(49): 15250–15255.

[62] Downey, G. and S. I. Feldman. 1996. Implications of rejection sensitivity for intimate relationships. Journal of Personality and Social Psychology 70(6): 1327.

[63] Maiolatesi, A. J., K. A. Clark and J. E. Pachankis. 2022. Rejection sensitivity across sex, sexual orientation, and age: Measurement invariance and latent mean differences. Psychological Assessment 34(5): 431.

[64] Ahlqvist, S., B. London and L. Rosenthal. 2013. Unstable identity compatibility: How gender rejection sensitivity undermines the success of women in science, technology, engineering, and mathematics fields. Psychological Science 24(9): 1644–1652.

[65] Morita, T., M. Asada and E. Naito. 2021. Gray-matter expansion of social brain networks in individuals high in public self-consciousness. Brain Sciences 11(3): 374.

[66] Early Social Interactions Build Connections in the Brain, Accessed on May 1st, 2025, https://headstart.gov/.

[67] Chaplin, T. M., K. L. Mauro and C. E. Niehaus. 2022. Effects of parenting environment on child and adolescent social-emotional brain function. Curr Top Behav Neurosci. 54: 341–372. doi: 10.1007/7854_2021_276. PMID: 34761364; PMCID: PMC10016201.

[68] Baumrind, D. 1967. Child care practices anteceding three patterns of preschool behavior. Genetic Psychology Monographs.

[69] Kopala-Sibley, D. C., M. Cyr, M. C. Finsaas, J. Orawe, A. Huang, N. Tottenham et al. 2020. Early childhood parenting predicts late childhood brain functional connectivity during emotion perception and reward processing. Child Development 91(1): 110–128.

[70] Tottenham, N. 2012. Human amygdala development in the absence of species-expected caregiving. Developmental Psychobiology 54(6): 598–611.

[71] Pozzi, E., J. G. Simmons, C. A. Bousman, N. Vijayakumar, K. O. Bray, O. Dandash et al. 2020. The influence of maternal parenting style on the neural correlates of emotion processing in children. Journal of the American Academy of Child and Adolescent Psychiatry 59(2): 274–282.

[72] Pittman, L. D. and A. Richmond. 2007. Academic and psychological functioning in late adolescence: The importance of school belonging. The Journal of Experimental Education 75(4): 270–290.

[73] Walton, G. M. and G. L. Cohen. 2007. A question of belonging: race, social fit, and achievement. Journal of Personality and Social Psychology 92(1): 82.

[74] Peters-Davis, N. and J. Shultz. 2015. Challenges of Multicultural Education: Teaching and Taking Diversity Courses. Routledge.

[75] Allen, K. A., M. L. Kern, C. S. Rozek, D. M. McInerney and G. M. Slavich. 2021. Belonging: A review of conceptual issues, an integrative framework, and directions for future research. Australian Journal of Psychology 73(1): 87–102.

[76] Ghorabi, F., N. Babakhani and F. Kashani. 2022. Determining the Effectiveness of Integration of Collaborative for Academic, Social, and Emotional Learning (CASEL) and Social Problem-Solving Skills on Internalizing and Externalizing Emotional and Behavioral Problems and Academic Performance of Fifth-Grade Primary School Students.

[77] Christidi, F., R. Migliaccio, H. Santamaría-García, G. Santangelo and F. Trojsi. 2018. Social cognition dysfunctions in neurodegenerative diseases: neuroanatomical correlates and clinical implications. Behavioural Neurology 2018(1): 1849794.

[78] Bales, R. F. 1950. Interaction process analysis; a method for the study of small groups.

[79] Zink, C. F., Y. Tong, Q. Chen, D. S. Bassett, J. L. Stein and A. Meyer-Lindenberg. 2008. Know your place: neural processing of social hierarchy in humans. Neuron 58(2): 273–283.

[80] Mason, M., J. C. Magee and S. T. Fiske. 2014. Neural substrates of social status inference: Roles of medial prefrontal cortex and superior temporal sulcus. Journal of Cognitive Neuroscience 26(5): 1131–1140.

[81] Feng, C., Z. Li, X. Feng, L. Wang, T. Tian and Y. J. Luo. 2016. Social hierarchy modulates neural responses of empathy for pain. Social Cognitive and Affective Neuroscience 11(3): 485–495.

[82] Williams, K. D., J. P. Forgas and W. Von Hippel. 2013. The Social Outcast: Ostracism, Social Exclusion, Rejection, and Bullying. Psychology Press.

[83] Kross, E., M. G. Berman, W. Mischel, E. E. Smith and T. D. Wager. 2011. Social rejection shares somatosensory representations with physical pain. Proceedings of the National Academy of Sciences 108(15): 6270–6275.

[84] Leary, M. R., A. L. Haupt, K. S. Strausser and J. T. Chokel. 1998. Calibrating the sociometer: The relationship between interpersonal appraisals and the state self-esteem. Journal of Personality and Social Psychology 74(5): 1290.

[85] Kross, E., T. Egner, K. Ochsner, J. Hirsch and G. Downey. 2007. Neural dynamics of rejection sensitivity. Journal of Cognitive Neuroscience 19(6): 945–956.

[86] Burklund, L. J., Eisenberger, N. I. and Lieberman, M. D. 2007. The face of rejection: rejection sensitivity moderates dorsal anterior cingulate activity to disapproving facial expressions. Social Neuroscience 2(3-4): 238–253.

[87] Gündel, H., M. F. O'Connor, L. Littrell, C. Fort and R. D. Lane. 2003. Functional neuroanatomy of grief: an FMRI study. American Journal of Psychiatry 160(11): 1946–1953.

[88] Davidson, R. J., K. M. Putnam and C. L. Larson. 2000. Dysfunction in the neural circuitry of emotion regulation—a possible prelude to violence. Science 289(5479): 591–594.

[89] Likhtik, E., J. G. Pelletier, R. Paz and D. Paré. 2005. Prefrontal control of the amygdala. Journal of Neuroscience 25(32): 7429–7437.

[90] Beauregard, M., J. Levesque and P. Bourgouin. 2001. Neural correlates of conscious self-regulation of emotion. The Journal of Neuroscience 21(18): RC165.

CHAPTER 4

Learning Mathematics and Brain

Emre Tokgoz,[1,*] *Percin Karakol*[2] and *Sophia Marra*[3]

4.1 Introduction

A wide range of implications are associated with learning and comprehension of mathematics at different ages of learning [1]. From a physiological standpoint, brain's elasticity is mentioned to be strongly tied to mathematics learning while cognitive functioning has also been significantly impacted by it as well [2]. A variety of areas are investigated to furthermore understand such effects and math-based conditions that may be faced by individuals that included but not limited to the following:

- Physical states of the brain
- Neurotransmitter activities
- Regional brain activations
- Cognitive skill development and brain chemistry
- Aging and neurogenesis

The mathematical complexity of the brain initiates from the differential behavior of cells and neurons depending on the physiological aspects of the brain that incorporates the speed, delays, and variations of neurons' synapsing infrastructures. In this sense, the time elapsed for a neural signal to be transmitted from a pre-synaptic neuron to post-synaptic neuron through a synapse is ranging anywhere

[1] School of Engineering and Technology, Farmingdale State College, New York, 11735 USA.
[2] Başakşehir Cam and Sakura City Hospital Plastic Reconstructive and Aesthetic Surgery, Istanbul, Turkiye.
[3] Frank H. Netter School of Medicine, Quinnipiac University, North Haven, CT, 06473, USA.
* Corresponding author: tokgoze@farmingdale.edu

DOI: 10.1201/9781003646044-4

between 0.5 to 1 millisecond that is known as synaptic delay, and this delay also depends on the synapse type that can be chemical or electrical synapsing. Fast acting neurotransmitters such as glutamate and gamma-aminobutyric acid are effective on the target cell within one millisecond through ion channels with the administration of the ligands. On the contrary, there are biochemically complicated processes such as the effects of biogenic amine and peptide neurotransmitters that may be completed over hundreds of milliseconds and even in minutes [3]. Noting such complexity of the systems, it is very likely that a closed-form mathematical formulation of brain-related operations cannot be determined by researchers as the biological variability is not only personal genetic dependent but also achieving another level of complexity with the illnesses and psychological factors that may impact the physiological functioning. The closest approaches yet achieved for understanding mathematics behind the brain appear as organization of the brain and the pathways with their designs. For instance, cerebral cortex is composed of modular processing units termed columns and the cortical areas specializing in some aspects of sensory, motor, cognitive, and limbic functioning abilities [4].

Noting such complexities, most of the research on mathematics of the brain focused on the ability of the brain to comprehend mathematics as well as the differences between individuals that are mathematically talented versus those who are not that may have disorder [5]. This difference, just like majority of the other brain-based research investigations, relied on functional MRI (f-MRI) and observations. In such considerations, critical aspects of considerations are included but are not limited to the following:

- Abstract mathematical thinking
- Mathematical logic processing
- Mathematical learning disability
- Mathematical processing ability
- Impact of mathematics on brain development
- Brain activities during mathematical and other concepts learning
- Specific mathematical conceptual processing of the brain such as digit processing
- Brain functioning of mathematical experts

There are certain brain regions that are determined to be activated during certain mathematical operations and the following are the corresponding regions that we will be covering in more detail in this chapter:

- Parietal lobe.
- Prefrontal cortex
- Frontal lobe
- Inferior parietal lobule
- Hemispheric regions
- Brain networks

The rest of this chapter focuses on two major aspects that link the above-mentioned investigations in the research literature: Brain regions playing significant roles in different mathematical concepts' learning and the neural network's functioning during mathematical learning. Impact of mathematics on thinking will also be covered from the psychological reflection of mathematics on thinking and learning.

4.2 Key Mathematical Foundations Playing a Role in Learning and Brain Development

There is tremendous amount of research effort and development focus on mathematical learning and brain. In this manner, there are certain dimensions that are brought into empirical observations including the following:

- Age
- Mathematical concepts
- Brain region activation
- Brain's functional processing
- Symbolic learning
- Theoretical mathematics information processing
- Learning disabilities

Mathematical learning has been characterized as an active and cumulative construction process as a result of extensive empirical evidence within a broad sociocultural context with the mathematical meaning, understanding, and skill development [45]. The essence of this constructivist approach is to build on prior knowledge through the demonstration of knowledge and skills acquired to learn more in an environment that requires physical and social interactions, and it has been applied to understand learning of children in and out of classroom [46]. This approach can have a positive impact on learning mathematics with the ability of constructing knowledge while it can also negatively impact in a way that may result in defective learning of mathematical procedure development and misconceptions that lead to misunderstandings. Prior knowledge misconstruction can result in knowledge gaps of individuals that may result in learning deficits of more advanced concepts.

In psychological research, one of the major focus areas is mathematics learning through studying general theories and approaches while thinking specification is ignored in such research [45]. The corresponding theories focused on differential cognitive development during mathematical learning. The dual nature of mathematics using behaviorism and connectionism highlighted the necessity of an in-depth approach to learning with the need of practice on learned information to have well defined information, associations, and skills developed [47].

Another approach that is developed for measuring mathematics learning process is by following the thinking mechanism of a mathematician. In this approach, inductive and deductive skills of a person are measured based on incomplete knowledge, and progressing in this approach requires the ability to handle tasks by developing solutions. For instance, the ability to convert real-world scenarios into abstraction and operate on the abstract form of the problem to be able to solve a real-world problem can be measured during mathematics learning process.

There are various levels of learners, and understanding mathematical comprehension of these different levels of learners based on different conceptual development is essential. The major areas of focus included primary, middle, and high school students' mathematics learning while university students' mathematical conceptual construction received limited attention. For instance, conceptual construction development of Science-Technology-Engineering-Mathematics (STEM) majors are investigated in [48–66] to have a better understanding of the university students' comprehension levels of the following. Action-Process-Object-Schema (APOS) theory is one of the theories used in these studies extensively to evaluate learners' understanding of conceptual construction.

- Numbers
- Asymptotes (horizontal and vertical) asymptotes
- Limits
- Functions
- Function graphs
- First and second derivatives
- Integrals
- Power series
- Function graphing
- Integration of concepts (such as derivative and power series)

One of the interesting results in these studies is identified to be the participants demonstration of second derivative understanding while not having the same level of understanding demonstration for the first derivative of functions.

There are other techniques utilized in relevant studies such as eye-trackers. For instance, additional empirical evidence is attained on undergraduate students devoting more of their attention to parts of proofs involving algebraic manipulation and less to logical statements when compared to expert mathematicians; an observation made by using eye-tracking techniques [68].

As a part of mathematical processing abilities, one can view different aspects of focus during mathematical learning [20]:

- Cognitive aspect focuses on how individuals think about mathematical constructs
- Affective aspects focusing on the enhancement and suppression of mental constructs [20].

There are various aspects of brain signaling that may occur based on different reactions to mathematical concepts learned. For instance, the limbic system is activated as a part of the emotions such as level of pleasure learning, tension, and anxiety that can have consequences such as developing sense of insecurity for learning mathematics and fear to solve mathematical problems.

4.3 Neural Networks and Mathematics

Several studies investigated the brain regions through neuroimaging with their roles in mathematical accomplishment abilities. In the case of basic and more advanced number processing operations, the areas of the brain network expand and get distributed across the parietal lobe [14] and even expands to include prefrontal cortex [21, 22] as well as other cortical and subcortical regions [9, 23, 24, 25]. The following are some of the key foundations in this area of research interest:

- Parietal lobe region has been identified to relate number processing and mathematical calculations in several studies [7, 8, 9, 10]. The following are some of the specifications related to mathematical operations and the corresponding brain regions taking place in such operations.
- Number processing magnitude has been observed to be associated with a specific region of the brain based on f-MRI studies. These locations included the bilateral horizontal segment of the intraparietal sulcus and the bilateral posterior superior parietal lobule as a region that extends from the posterior segment of the intraparietal sulcus to the precuneus [11]. The same arithmetic factors are determined to be regulated in these two regions based on blood oxygen observations and the characteristic of the arithmetic operation that included the following:
    - Problem's size [12]
    - Number of operands [13]
    - Strategy [14, 15, 16]
- The bilateral horizontal segment of the intraparietal sulcus is determined to be the only location that was selectively activated during the subtraction mathematical operation when compared to manual, visuospatial or linguistic tasks [17].
- Intraparietal sulcus has been associated with both symbolic and non-symbolic number processing tasks [19, 20].
- Diving into more details of specific parietal lobe regions and their roles in mathematical calculations, an f-MRI study is conducted focusing on observing changes based on subtraction and multiplication indicated the active role of the horizontal segment of the intraparietal sulcus during solving operations that require memory retrieval such as multiplication. The posterior superior parietal lobule appeared to take place in processing of basic subtraction and multiplication problems [6].
- The gray matter density in the right superior parietal lobule gyrus has been identified to be more in mathematic experts when compared to non-mathematical experts as a result of a whole-brain analysis while the grey matter was lower in the right intraparietal sulcus and left inferior frontal gyrus [18]. The behavior

performance was identified to be correlated with the grey matter density in the left superior frontal gyrus.

- Mathematics academicians evaluate statements from a variety of perspectives that are determined to result in increased network activation across a network of bilateral frontal, intraparietal and ventrolateral temporal regions while non-mathematics academicians are identified to not have the same characteristics [26].

Figure 1 is a high-level sketch of the Intraparietal Sulcus (IPS) and Prefrontal Cortex regions on the brain.

In addition to the research on the involvement of the cortical networks in mathematical expertise and individual differences, various white matter areas that connect neural network nodes have also been investigated, and fractional anisotropy is used as the common measure for understanding preservation levels of microstructural integrity to evaluate their role in such mathematical functioning activities [27].

One other area of investigation included the memory associated with the mathematical operation abilities. Visual imaging capabilities, long-term memory, and working memory are examples of domain-general cognitive processing operations that are observed to be correlated with mathematical ability [28]. Such supportive functions are studied to identify the numerical competency [31, 32, 33, 34, 35, 36] and numerical deficiencies [29, 30].

Processing of symbolic fractions and fraction words are identified to be associated with the anterior IPS activities in a study that was conducted on 12 participants [37]. For instance, mathematical fractions of the same magnitude in the fraction were presented to the participants initially (such as 1/5, 2/12) and the corresponding representation changed to symbolic and word representations of the fractions such as 1:2 and one half, respectively. In another study by the same authors [38], the authors studied non-symbolic proportion processing that included line proportion and dot proportions. The strongest effect is identified to occur in bilateral anterior IPS. Some other regions that are activated included bilateral prefrontal and precentral regions with mainly right lateral portion activations. It has been observed that mathematical fraction and proportion activation in IPS was changing as a function of the distance between two fractions. Another study that conducted the impact of fraction magnitude changes on brain activation was conducted to determine the difference between mathematical fractions' comparisons. The IPS was identified to react to such comparison tasks in a holistic fashion. IPS activation differential between rational and natural number processing of the brain was studied in [39]. Numbers with variational magnitudes in digit representations of numbers were presented to 16 adults. The results indicated the involvement of IPS in all magnitudes of number processing, only fractions activated a distinct subarea within IPS. There are numerous other studies that are conducted on fraction processing of the brain with f-MRI and EEG signaling observations and IPS is determined to play a dominant role in the functioning of the brain [40].

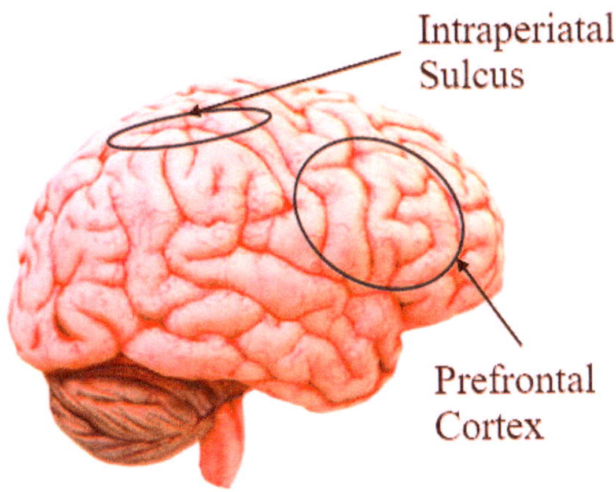

Fig. 1. Intraparietal Sulcus (IPS) and Prefrontal Cortex regions on the brain.

In addition to the above-mentioned ordinary mathematical considerations for understanding the brain's role, a variety of other mathematical calculation types have been investigated in research that included the following:

- **Approximate calculations.** The mean fractional anisotropy in the left anterior portion of superior longitudinal fasciculus of healthy individuals is determined to correlate with their mental approximate calculations upon a tractography application [41].
- **Behavioral and fNIRS results on Exact vs. Approximate Arithmetic.** In a study, exact arithmetic reaction time is observed to result in significantly longer time on approximate arithmetic exhibition depending on the problem size and operand processing duration during arithmetic calculations.
    - Approximate calculations activated the right region of the brain while exact arithmetic calculations are determined to correlate with left frontal gyrus region activation [42].
    - The right pre-motor cortex and supplementary motor cortex are determined to be actively functioning during approximation calculations.
    - Calculations using larger operands appeared to have activation on the
        - right dorsolateral prefrontal cortex,
        - right superior frontal gyrus,
        - pre- and post-supplementary motor cortex.

Hence the corresponding findings pointed to the idea of the left hemisphere of the brain to be active during exact arithmetic calculations while right portion of the brain being functional during approximate calculations suggesting a neural dissociation between exact and approximate arithmetic operation processing of the brain.

4.4 Impact of Mathematics on Thinking

There is a significant impact of mathematics on thinking and learning. The baseline of developed cognitive skills directly correlating positively with problem-solving, critical thinking, and reasoning. The learning of mathematics enhances brain functioning tremendously that can allow advanced and flexible thinking with working memory activation and attention to details. The following are some of the details of how advancing mathematical ability can impact thinking and learning:

- **Cognitive skill improvements.** Improvements in cognitive ability can advance as mathematical thinking advances. The associated areas of skill improvement can include logical reasoning, critical thinking, and ability to solve problems.
- **Brain functioning enhancement.** One of the physiological impacts of learning mathematics is establishing brain elasticity that also impacts cognitive functioning and abilities that lead to better established overall development with more advanced ability of brain functioning [43].
- **Flexible thinking development.** Mathematical problem solving requires creativity and approaching the solutions of a problem by using a variety of methods and approaches; hence, mathematical learning fosters resourcefulness, creativity, and flexible thinking abilities.
- **Working memory improvement and attention.** Memory and attention can be improved during mathematical task handling as a part of spatial skill improvement.
- **Mathematical anxiety reduction.** Mathematics is a natural part of daily life (such as shopping) and many people deal with mathematics either during their educational environment or work. This can cause anxiety, and such anxiety can be reduced by improving mathematical skills.
- **Embodied cognition changes.** Actions and perceptions interact to support learning during embodied cognition and mathematical thinking is a part of it. Mathematical learning can be influential on embodied cognition. For instance, pointing to the geometric representation of a graph and its algebraic equation to explain the concept can be supportive in learning mathematics and brain development of both geometric and algebraic function representation comprehension [44].
- **Culture of Mathematical Thinking.** Applying mathematical thinking in real life to make it a habit can result in a systematic way of living with definitions, abstractions, and establishing connections incorporated into real-life comprehension.
- **Expansion of learning ability.** Mathematical understanding of various concepts are required to be comprehended for advancement of learning in many other fields and such advancement is the prerequisite to advance in many other fields.

A long-term mathematical thinking ability is summarized into three stages based on the work of [41]:

- **1st Stage.** Ability to advance mathematics starts with practical mathematical experiences that are based on spatial awareness; The operational symbolism initiates in this first step of conceptual pre-processing allows continuation with understanding more advanced mathematics. Development of symbolic mathematical representations leads to abstraction of mathematics with likely impact on brain functioning capabilities.
- **2nd Stage.** Led by perception of mathematical conceptualization and operational experiences with symbols, the second stage allows advancement to theoretical mathematics that rely on definitions and mathematical deductions. In this stage, the ability to conceptualize prior knowledge furthermore advancements with mathematical deductions. Continued advancements in symbolic and the corresponding operational comprehension yields to mathematical deductions from the properties of known operations on newer operations that are more advanced.
- **3rd Stage.** The top level of advancement is achieved due to the ability to learn and apply developed axiomatic approaches in a formal manner based on the definitions and set-theoretical axioms after deducing their conceptual meaning.

Figure 2 is a visual representation of the corresponding three stages.

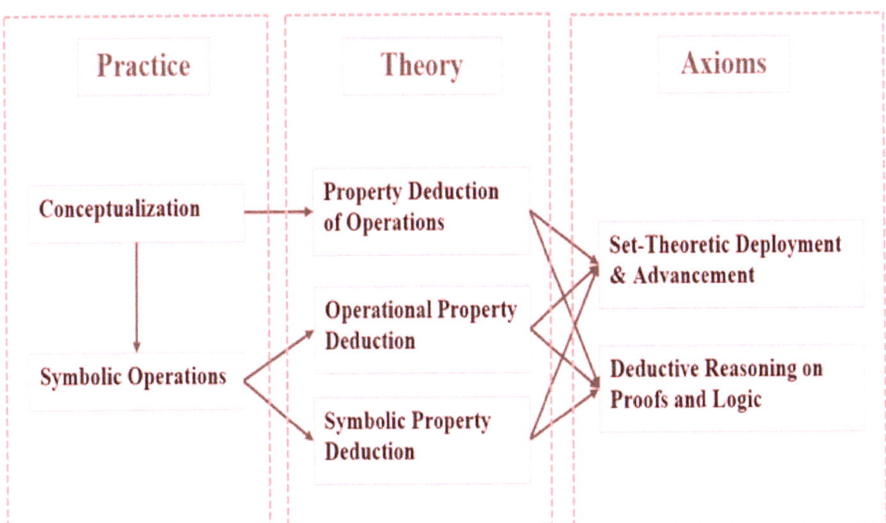

Fig. 2. A broad overview of the three stages of mathematical learning in [41].

References

[1] Zacharopoulos, G., F. Sella and R. Cohen Kadosh. 2021. The impact of a lack of mathematical education on brain development and future attainment. Proceedings of the National Academy of Sciences 118(24): e2013155118.

[2] Beddington, J., C. L. Cooper, J. Field, U. Goswami, F. A. Huppert, R. Jenkins et al. 2008. The mental wealth of nations. Nature 455(7216): 1057–1060.

[3] Greengard, P. 2001. The neurobiology of slow synaptic transmission. Science 2; 294(5544): 1024–30. doi: 10.1126/science.294.5544.1024. PMID: 11691979.

[4] Roe, A. W. 2019. Columnar connectome: toward a mathematics of brain function. Network Neuroscience 3(3): 779–791.

[5] Wirebring, L. K., J. Lithner, B. Jonsson, Y. Liljekvist, M. Norqvist and L. Nyberg. 2015. Learning mathematics without a suggested solution method: Durable effects on performance and brain activity. Trends in Neuroscience and Education 4(1-2): 6–14.

[6] Andres, M., B. Pelgrims, N. Michaux, E. Olivier and M. Pesenti. 2011. Role of distinct parietal areas in arithmetic: an fMRI-guided TMS study. Neuroimage 54(4): 3048–3056.

[7] Cipolotti, L. and A. de Lacy Costello. 1995. Selective impairment for simple division. Cortex 31(3): 433–449.

[8] Dehaene, S. and L. Cohen. 1991. Two mental calculation systems: A case study of severe acalculia with preserved approximation. Neuropsychologia 29(11): 1045–1074.

[9] Dehaene, S. and L. Cohen. 1997. Cerebral pathways for calculation: Double dissociation between rote verbal and quantitative knowledge of arithmetic. Cortex 33(2): 219–250.

[10] Gerstmann, J. 1940. Syndrome of finger agnosia, disorientation for right and left, agraphia and acalculia: Local diagnostic value. Archives of Neurology and Psychiatry 44(2): 398–408.

[11] Dehaene, S., M. Piazza, P. Pinel and L. Cohen. 2005. Three parietal circuits for number processing. pp. 433–453. In The Handbook of Mathematical Cognition. Psychology Press.

[12] Stanescu-Cosson, R., P. Pinel, P. F. van de Moortele, D. Le Bihan, L. Cohen and S. Dehaene. 2000. Understanding dissociations in dyscalculia: a brain imaging study of the impact of number size on the cerebral networks for exact and approximate calculation. Brain 123(11): 2240–2255.

[13] Menon, V., S. M. Rivera, C. D. White, G. H. Glover and A. L. Reiss. 2000. Dissociating prefrontal and parietal cortex activation during arithmetic processing. Neuroimage 12(4): 357–365.

[14] Delazer, M., F. Domahs, L. Bartha, C. Brenneis, A. Lochy, T. Trieb et al. 2003. Learning complex arithmetic—an fMRI study. Cognitive Brain Research 18(1): 76–88.

[15] Delazer, M., A. Ischebeck, F. Domahs, L. Zamarian, F. Koppelstaetter, C. M. Siedentopf et al. 2005. Learning by strategies and learning by drill—evidence from an fMRI study. Neuroimage 25(3): 838–849.

[16] Ischebeck, A., L. Zamarian, C. Siedentopf, F. Koppelstätter, T. Benke, S. Felber et al. 2006. How specifically do we learn? Imaging the learning of multiplication and subtraction. Neuroimage 30(4): 1365–1375.

[17] Simon, O., J. F. Mangin, L. Cohen, D. Le Bihan and S. Dehaene. 2002. Topographical layout of hand, eye, calculation, and language-related areas in the human parietal lobe. Neuron 33(3): 475–487.

[18] Popescu, T., E. Sader, M. Schaer, A. Thomas, D. B. Terhune, A. Dowker et al. 2019. The brain-structural correlates of mathematical expertise. Cortex 114: 140–150.

[19] Harvey, B. M., B. P. Klein, N. Petridou and S. O. Dumoulin. 2013. Topographic representation of numerosity in the human parietal cortex. Science 341(6150): 1123–1126.

[20] Tall, D. 2019. From biological brain to mathematical mind: The long-term evolution of mathematical thinking. Interdisciplinary Perspectives on Math Cognition 1–28.

[21] Kaufmann, L., G. Wood, O. Rubinsten and A. Henik. 2011. Meta-analyses of developmental fMRI studies investigating typical and atypical trajectories of number processing and calculation. Developmental Neuropsychology 36(6): 763–787.

[22] Nieder, A. and S. Dehaene. 2009. Representation of number in the brain. Annual Review of Neuroscience 32(1): 185–208.

[23] Hittmair-Delazer, M., C. Semenza and G. Denes. 1994. Concepts and facts in calculation. Brain 117(4): 715–728.

[24] Menon, V., R. C. Kadosh and A. Dowker. 2015. The Oxford Handbook of Numerical Cognition.

[25] Qin, S., S. Cho, T. Chen, M. Rosenberg-Lee, D. C. Geary and V. Menon. 2014. Hippocampal-neocortical functional reorganization underlies children's cognitive development. Nature Neuroscience 17(9): 1263–1269.

[26] Amalric, M. and S. Dehaene. 2016. Origins of the brain networks for advanced mathematics in expert mathematicians. Proceedings of the National Academy of Sciences 113(18): 4909–4917. doi: 10.1073/pnas.1603205113.

[27] Moeller, K., K. Willmes and E. Klein. 2015. A review on functional and structural brain connectivity in numerical cognition. Frontiers in Human Neuroscience 9: 227.

[28] Giorello, G. and C. Sinigaglia. 2005. Oliver Heaviside's "Dinner" Algebraic Imagination and Geometrical Rigour. pp. 11–32. In The Role of Mathematics in Physical Sciences: Interdisciplinary and Philosophical Aspects. Springer Netherlands.

[29] Passolunghi, M. C. and L. S. Siegel. 2004. Working memory and access to numerical information in children with disability in mathematics. Journal of Experimental Child Psychology 88(4): 348–367.

[30] Wilson, K. M. and H. L. Swanson. 2001. Are mathematics disabilities due to a domain-general or a domain-specific working memory deficit? Journal of Learning Disabilities 34(3): 237–248.

[31] Bull, R., K. A. Espy and S. A. Wiebe. 2008. Short-term memory, working memory, and executive functioning in preschoolers: Longitudinal predictors of mathematical achievement at age 7 years. Developmental Neuropsychology 33(3): 205–228.

[32] Dumontheil, I. and T. Klingberg. 2012. Brain activity during a visuospatial working memory task predicts arithmetical performance 2 years later. Cerebral Cortex 22(5): 1078–1085. doi: 10.1093/cercor/bhr175.

[33] O'Boyle, M. W. 2008. Mathematically gifted children: Developmental brain characteristics and their prognosis for well-being. Roeper Review 30(3): 181–186.

[34] Pesenti, M., L. Zago, F. Crivello, E. Mellet, D. Samson and B. Duroux. 2001. Mental calculation in a prodigy is sustained by right prefrontal and medial temporal areas. Nature Neuroscience 4(1): 103–108. doi: 10.1038/82831.

[35] Thompson, J. M., H. C. Nuerk, K. Moeller and R. C. Kadosh. 2013. The link between mental rotation ability and basic numerical representations. Acta Psychologica 144(2): 324–331.

[36] van Dijck, J. P. and W. Fias. 2011. A working memory account for spatial–numerical associations. Cognition 119(1): 114–119.

[37] Jacob, S. N. and A. Nieder. 2009. Notation-independent representation of fractions in the human parietal cortex. Journal of Neuroscience 29(14): 4652–4657.

[38] Jacob, S. N. and A. Nieder. 2009. Tuning to non-symbolic proportions in the human frontoparietal cortex. European Journal of Neuroscience 30(7): 1432–1442.

[39] DeWolf, M., J. N. Chiang, M. Bassok, K. J. Holyoak and M. M. Monti. 2016. Neural representations of magnitude for natural and rational numbers. NeuroImage 141: 304–312.

[40] Wortha, S. M., A. Obersteiner and T. Dresler. 2022. Neurocognitive foundations of fraction processing. pp. 289–315. In Handbook of Cognitive Mathematics. Cham: Springer International Publishing.

[41] Tsang, J. M., R. F. Dougherty, G. K. Deutsch, B. A. Wandell and M. Ben-Shachar. 2009. Frontoparietal white matter diffusion properties predict mental arithmetic skills in children. Proceedings of the National Academy of Sciences 106(52): 22546–22551.

[42] Yue, T., B. Guan and Y. Wu. 2025. Cognitive and neural differences in exact and approximate arithmetic using the production paradigm: An fNIRS study. Behavioral Sciences 15(1): 33.

[43] Zhang, Y., S. Liu and X. Shang. 2021. An MRI study on effects of math education on brain development using multi-instance contrastive learning. Frontiers in Psychology 12: 765754.

[44] Tran, C., B. Smith and M. Buschkuehl. 2017. Support of mathematical thinking through embodied cognition: Nondigital and digital approaches. Cognitive Research: Principles and Implications 2: 1–18.
[45] De Corte, E., B. Greer and L. Verschaffel. 1996. Mathematics teaching and learning. Handbook of Educational Psychology 1: 491–549.
[46] Ginsburg, H. P. and S. L. Golbeck. 2004. Thoughts on the future of research on mathematics and science learning and education. Early Childhood Research Quarterly 19(1): 190–200.
[47] Maggioni, L. and P. A. Alexander. 2010. Knowledge domains and domain learning.
[48] Tokgöz, E. 2024. Evaluation of Mathematical Building Blocks Impacting STEM Majors' Ability to Solve Conceptual Power Series Questions, ASEE 2024 Middle Atlantic Section Conference Proceedings, Article ID #: 45311.
[49] Tokgöz, E. 2024. Understanding STEM Students' Conceptual Differentiability Knowledge Through Analysis of Sub-concept Cognition, ASEE 2024 Middle Atlantic Section Conference Proceedings, Article ID #: 45312.
[50] Tokgöz, E., E. N. Tekalp, S. B. Tekalp, H. Tekalp, S. E. Scarpinella and M. Giannone. 2022. Analysis of STEM students accumulating calculus knowledge to graph a function. Proceedings of 2022 ASEE Annual Conference and Exposition, Minneapolis, MN.
[51] Tokgöz, E., E. N. Tekalp, S. B. Tekalp, H. Tekalp, S. E. Scarpinella and M. Giannone. 2022. STEM Students' Technology Choice Differences for Solving a Graphing Question - Two Different Institution Students' Preferences for Solving a Graphing Question, Proceedings of 2022 ASEE Annual Conference and Exposition, Minneapolis, MN.
[52] Tokgöz, E., E. N. Tekalp, S. B. Tekalp, H. Tekalp, S. E. Scarpinella and M. Giannone. 2022. An analysis of STEM students' integral and area under the curve knowledge. Proceedings of 2022 ASEE Annual Conference and Exposition, Minneapolis, MN.
[53] Tokgöz, E., S. E. Scarpinella and M. Giannone. 2021, July. Analysis of STEM Students' Ability to Respond to Algebra, Derivative, and Limit Questions for Graphing a Function, 2021 ASEE Annual Conference. https://peer.asee.org/36685.
[54] Tokgöz, E., E. N. Tekalp, S. B. Tekalp and H. A. Tekalp. 2021, July. Undergraduate STEM Students' Comprehension of Function Series and Related Calculus Concepts. 2021 ASEE Annual Conference, https://peer.asee.org/37950.
[55] Tokgöz, E., S. E. Scarpinella and M. Giannone. 2021, July. Technology Decisions of Engineering Students for Solving Calculus Questions 2021 ASEE Annual Conference, https://peer.asee.org/37830.
[56] Inglis, M. and L. Alcock. 2012. Expert and novice approaches to reading mathematical proofs. Journal for Research in Mathematics Education 43(4): 358–390.
[57] Tokgöz, E., E. N. Tekalp, S. B. Tekalp and H. A. Tekalp. 2020. Undergraduate STEM Students' Role in Making Technology Decisions for Solving Calculus Questions and the Impact of These Decisions on Learning Calculus, 127th Annual ASEE Conference Proceedings, Computer Science Division, paper ID # 29730, Montreal, Canada, 2020.
[58] Tokgöz, E., S. B. Tekalp, E. N. Tekalp and H. A. Tekalp. 2020. Qualitative and Quantitative Analysis of University Students' Ability to Relate Calculus Knowledge to Function Graphs, 127th Annual ASEE Conference Proceedings, Mathematics Division, paper ID # 29726, Montreal, Canada, 2020.
[59] Tokgöz, E. 2019. STEM Majors' Ability to Calculate Taylor Series' Derivative and Integral, ASEE Annual Conference Proceedings – Mathematics Division, paper ID # 25091.
[60] Tokgöz, E. and H. Ceyhan. 2018. University Students Ability to Interconnect the Calculus Concepts and Function Graphing, ASEE Annual Conference Proceedings – Mathematics Division, paper ID # 21382.
[61] Tokgöz, E. 2018. Conceptual Power Series Knowledge of STEM Majors, ASEE Annual Conference Proceedings – Mathematics Division, paper ID # 21246.
[62] Tokgöz, E. and H. Ceyhan. 2017. Integral Concept and Decision Making: Do the STEM Majors Know When to Use Numerical Methods for Integral Approximation? ASEE Annual Conference Proceedings, paper ID # 17833.
[63] Tokgöz, E. 2016. Evaluation of Engineering and Mathematics Majors' Riemann Integral Definition Knowledge by Using APOS Theory, ASEE Annual Conference Proceedings, paper ID# 14461, 2016.

[64] Tokgöz, E. 2016. STEM Majors' Ability to Relate Integral and Area Concepts, ASEE Annual Conference Proceedings, paper ID # 14462.

[65] Tokgöz, E. and G. C. Gualpa. 2015. STEM Majors' Cognitive Calculus Ability to Sketch a Function Graph, ASEE Annual Meeting Proceedings, Mathematics Division, Seattle, Washington, Paper ID # 12661, 2015.

[66] Tokgöz, E. 2015. Analysis of STEM Majors' Calculus Knowledge by Using APOS Theory on a Quotient Function Graphing Problem. ASEE Annual Meeting Proceedings, Mathematics Division, Seattle, Washington, Paper ID # 12664.

Chapter 5

Alzheimer's Disease

Sophia Marra,[1,*] *Emre Tokgoz*[2] and *Percin Karakol*[3]

5.1 Introduction

Alzheimer's Disease (AD) is one of the most common neurodegenerative diseases, affecting up to 416 million people worldwide. This means that 22% of people over the age of 50 years old fall on the AD spectrum. This spectrum ranges from AD with dementia, prodromal AD and preclinical AD which each account for 32, 69, 315 million patients, respectively [70]. As the median age increases, the number of patients with AD is expected to triple in the US by the year 2050 [71]. Each year approximately $1 trillion US dollars are spent on dementia worldwide, making it a significant economic burden globally [54].

AD is characterized by the development of amyloid-beta plaques, tau neurofibrillary tangles, and neurodegeneration. This diagnosis can be life altering with symptoms including cognitive decline, memory loss, decrease in executive function and trouble carrying out daily tasks. Late-onset Alzheimer's Disease (LOAD) describes development of AD after the age of 65 years old. LOAD accounts for approximately 90–95% of all AD cases. LOAD pathogenesis is multifactorial and usually sporadic. In comparison, early-onset Alzheimer's Disease is only about 5% of all AD cases. It is almost always familial in origin and due to inheritance of a genetic mutation [69].

[1] Frank H. Netter School of Medicine, Quinnipiac University, North Haven, CT, 06473, USA.
[2] School of Engineering and Technology, Farmingdale State College, New York, 11735 USA.
[3] Başakşehir Cam and Sakura City Hospital Plastic Reconstructive and Aesthetic Surgery, Istanbul, Turkiye.
* Corresponding author: Sophia.Marra@quinnipiac.edu

DOI: 10.1201/9781003646044-5

5.2 Presentation/Diagnosis

5.2.1 Stages of Progression/Clinical Presentation

The most up to date understanding of Alzheimer's Disease describes the disease on a continuum starting with the preclinical stage and ending with severe AD dementia. The first neuropathological changes can occur many years prior to one's first clinical symptoms [19]. Various institutional bodies and working groups define the stages of AD using different terminology, but they all follow a similar progression as seen in Figure 1. The Food and Drug Administration describes AD progression from Stage 1 to Stage 6 (Figure 1 top row). The International Working Group and National Institute on Aging-Alzheimer's Association share similar staging criteria starting with presymptomatic/preclinical patients (Stage 1) which in will progress to prodromal AD with mild cognitive impairment (Stage 2) in about 20% of patients [25]. 10% of patients with mild cognitive impairment will go on developing dementia [1]. Dementia in AD patients is further subdivided into mild, moderate and severe (Stages 4, 5, and 6 respectively).

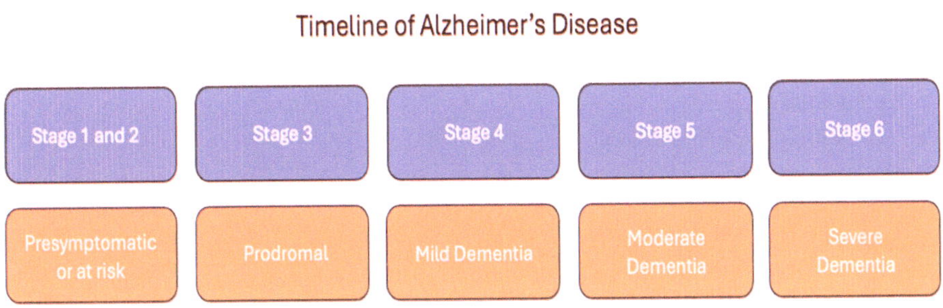

Fig. 1. Stages of Alzheimer's Progression [17].

5.2.1.1 Preclinical Stage

The ability to identify patients in the preclinical stage, even before their very first clinical symptom, is therapeutically relevant to enact preventative and early intervention treatment. The preclinical stage is commonly termed the "silent" stage as a patient could unknowingly be in this phase for decades [38]. The current definition for the preclinical stage is from the first pathological brain changes to a patient's first clinical symptom of AD [18]. But in practice, defining the initial stages of AD is quite debated.

Alzheimer's disease was at one point diagnosed solely on clinical presentation and confirmed via postmortem analysis [22]. Current research strives to understand what the most appropriate metrics are to define the beginning stages of AD. Some believe AD begins with the first cognitive decline symptoms whereas others recognize AD starting with the presence of biomarkers like amyloid-beta plaques and tau tangles [18]. Defining AD using these metrics is complicated by the fact that some individuals with AD biomarkers do not go on to have clinical presentations of the disease. Low levels of amyloid and tau proteins can be seen in the aging population that do not go onto develop AD [20].

The preclinical stage is subdivided into three stages [23]:

1. Amyloid-beta and tau protein accumulation in the neocortex and entorhinal cortex [28]
2. Presence of amyloid-beta plaque buildup with neurodegeneration
3. Amyloid-beta plaques, neurodegeneration and subtle cognitive decline not yet reaching threshold for diagnosis of Mild Cognitive Impairment

5.2.1.2 Mild Cognitive Impairment Stage

Mild Cognitive Impairment (MCI) is the next stage of AD where patients start to display clinical symptoms. This is considered a type of "intermediate zone" where individuals display cognitive changes but do not yet meet the criteria for dementia [29].

Cognitive function is subdivided into 5 domains [30]:

1. Learning and Memory
2. Language
3. Visuospatial
4. Executive
5. Psychomotor

Diagnosis of MCI indicates deficits in one of these five areas, whereas dementia suggests deficits in more than one domain.

Episodic short-term memory loss is the most common first symptom reported by AD patients [1]. This could include recent conversations, names, or places [24]. At this point patients' long-term memory is likely still intact. Patients may also experience non-memory related impairments including executive function and language. Executive function changes can include familiar activities of daily life such as driving, cooking, and maintaining one's finances [1]. Patients with MCI due to AD may exhibit linguistic challenges such as struggling to find the right word in conversation. At this point, patients still retain independence in completing daily tasks [24].

5.2.1.3 Dementia- Mild, Moderate, Severe

The percentage of those patients with MCI that develop dementia is debated. Studies report that it may be between 32.7%–70% of individuals [26, 27]. Memory changes may be accompanied by linguistic deficits including trouble finding words (anomia) resulting in circumlocution, misusing words, and an overall difficulty effectively partaking in spontaneous conversation [1].

Additionally, patients may have difficulty completing physical tasks even when the command is understood, and the necessary muscles groups are functional (apraxia) [1]. A decline in visuospatial abilities may cause patients to get lost in familiar surroundings [17]. Delusions occur in 20–40% of individuals with AD. Visual hallucinations are the most reported, but auditory and olfactory hallucinations are also possible. Patients also commonly have trouble sleeping resulting in fragmented sleep patterns. Additionally, up to 50% of individuals with AD may exhibit disruptive behaviors [1]. These symptoms may first appear in the mild dementia stage, but the severity intensifies throughout the disease progression with symptoms highest in the later stages [31].

While some of the symptoms among mild, moderate and severe dementia may be similar, it is the severity that differentiates the three. Patients with moderate to severe dementia most often require substantial care and medical support [30].

5.2.2 Pathophysiology

Dr. Alois Alzheimer first described the presence of amyloid plaques and hyperphosphorylated tau neurofibrillary tangles in the brains of people with Alzheimer's postmortem in 1907 [33]. While the exact cause of Alzheimer's is still debated, the disease is characterized by the presence of three hallmark features:

1. **Neuritic plaques.** Amyloid-beta plaques are the result of the improper proteolytic splicing of amyloid precursor protein (APP), a transmembrane protein coded on the twenty-first chromosome. The improper slicing results in a conversion of α-helical sheets to β-sheets, initiating oligomerization and the formation of insoluble amyloid plaques called senile plaques [34]. Throughout the progression of AD, these senile plaques will collect neural debris and become entangled in enlarged axon endings to create neuritic plaques [16, 39].
2. **Intracellular neurofibrillary tangles.** Tau protein plays an important role in the stability and formation of axonal microtubules, one of the key players in intracellular transport [16]. Current findings suggest that amyloid-beta plaque buildup triggers the hyperphosphorylation of tau protein intracellularly via several kinases [39]. This hyperphosphorylation results in the misfolding and aggregation of tau proteins into fibrils. Fibrils then further aggregate to create a paired helical structure called neurofibrillary tangles (NFTs). NFTs first form in the hippocampus in AD and then spread throughout the cerebral cortex [40].
3. **Neuronal Degeneration.** Postmortem studies show that there is significant reduction in brain volume and number of neurons in patients with AD [41, 42]. Additionally, researchers have shown a > 50% hippocampal and cerebral cortex neuron decrease in symptomatic AD patients [43, 44]. The literature suggests that it is the granulovacuolar degeneration (GVD) of the hippocampal pyramidal neurons that is most common [16, 45]. GVD results in large, cytoplasmic, double-membrane vesicles within neurons [46]. Specifically, this degeneration seems to impact laminae III and IV the most. It is this degeneration and loss of neuronal density that appears to be more closely implicated in the cognitive decline seen in AD, compared to the accumulation of plaques [16].

While the major hallmarks of Alzheimer's Disease (listed above) have been well-documented, the exact mechanism by which AD occurs is still debated. Current literature suggests the following major hypotheses for the development of AD:

1. **Amyloid Hypothesis.** Current literature suggests that this is a likely mechanism by which AD develops. This hypothesis suggests that it is the improper splicing of APP, resulting in an accumulation of amyloid-beta plaques in an "amyloid cascade" that causes AD. This cascade also triggers the development of tau tangles. Some researchers believe this to be an incomplete explanation because many therapies targeting this cascade have not been successful [53]. Additionally, cognitively healthy individuals have been found to have amyloid deposits and not show any symptoms of AD [55].
2. **Cholinergic Hypothesis.** This hypothesis is based on the significant cholinergic neuronal loss early in the development of AD. Specifically, cholinergic neurons in the Nucleus Basalis of Meynert, with projections to the hippocampus and entorhinal cortex, seem to be damaged. This neuronal loss is seen even before degeneration of the hippocampus and entorhinal cortex and is subsequently the topic of current research to investigate its usefulness as a biomarker and drug target [51]. This hypothesis underscores the importance of cholinergic neurons on cognitive processing, a claim bolstered by the negative effects anticholinergics have been shown to have on elderly individuals' memory [52]. Additionally, current literature suggests that amyloid-beta causes a decrease in cholinergic synapses and dysfunctional Ach release from vesicles [1].
3. **Metabolism.** Diabetes Mellitus (DM) is currently the most common metabolic disorder. It is characterized by either insensitivity to insulin or the lack of insulin secretion. Current literature suggests that diabetes may lead to mild cognitive impairments and eventually AD [56]. Although DM has been linked to AD, the exact mechanism is still being researched. One proposed mechanism highlights the link between APP and insulin receptor, both of which are cleaved via beta-secretase [57].

   Obesity has also been implicated with AD. Obesity is defined as a BMI greater or equal to 30 kg/m^2 [58]. Specifically, studies have shown that increased visceral adipose tissue (VAT) has been associated with increased neuronal degeneration and cognitive decline [60]. Additionally, increased VAT metabolism has been correlated with increased amyloid-beta levels [59].
4. **Inflammation.** Current research suggests that chronic inflammation can result in decreased integrity of the blood brain barrier (BBB) leading to the development of AD. The BBB allows necessary nutrients to enter the brain and facilitates the removal/blockage of harmful toxins and metabolite from the brain. The proposed mechanism by which inflammation can lead to AD development starts with the activation of inflammatory mediators and immune cells in response to a perceived pathogen. Chronic activation of these inflammatory mediators can result in the destruction of the BBB. This in turn leads to increase permeability of toxins and inflammatory molecules into the brain resulting in AD development [2].

5.2.3 Biomarkers

Previously, AD was diagnosed solely based on clinical symptoms and only confirmed postmortem by confirming the presence and quantity of amyloid-beta and tau [22]. Current research over the past decade has made great strides to develop cerebral spinal fluid (CSF) and blood biomarker test to detect the presence of AD pathological products [50].

CSF methods are a more established practice clinically compared to blood-based biomarkers. Recent strides have been designed to investigate the efficacy of blood biomarkers compared to CSF. This major push towards blood-based biomarkers is driven by the potential cost, time, manpower that could be saved using this technique.

The current research literature suggests investigating AD via biomarker using the "ATN classification criteria" which stands for "Amyloid-beta", "tau", and "neurodegenerations". As mentioned previously, these are the three widely accepted characteristics of AD pathology [16].

The following are the current CSF biomarkers available to investigate AD pathology:

1. Amyloid-beta (Aβ42, Aβ42/Aβ40 ratio)
2. Tau protein (phosphorylated-tau 217, non-phosphorylated tau, total tau)
3. Neurodegeneration (neurofilament light chain protein)

Out of all the measures highlighted, it is currently standard practice to measure CSF Aβ42 levels clinically. The following three AD hallmarks can be examined via the blood-based biomarkers listed below [61]:

1. Amyloid-beta (Aβ42/Aβ40 ratio)
2. Phosphorylated tau (p-tau181, p-tau217, p-tau231)
3. Neurodegeneration (neurofilament light chain protein)

Amyloid-beta and phosphorylated-tau biomarkers are analyzed via mass spectrometry assays and immunoassays. Neurofilament light chain protein (NfL) has become one of the most common markers of neurodegeneration. Specifically, it indicates the presence of neuroaxonal injury. One consideration that must be made when implementing this test is that the presence of NfL is not specific to AD. Elevated NfL levels are also associated with amyotrophic lateral sclerosis (ALS), frontotemporal dementia, traumatic brain injury, peripheral neuropathy and cardiac arrest. Recent literature points to the usefulness of a newly developed brain-derived tau assay that would assess AD-specific neurodegeneration [62].

Some studies also suggest glial fibrillary acidic protein (GFAP) as an important blood-based biomarker of AD. GFAP indicates the presence of astrocytic activation, commonly reported in response to amyloid-beta pathology in AD. Like NfL, GFAP is not AD-specific but the level of GFAP increase in AD versus non-AD diseases is relatively large [63, 64].

One of the major benefits of both CSF and blood-based biomarkers is that it allows for the visualization of AD pathology even before obvious clinical symptoms. The downside is that it does not give information about the severity of AD [1].

Another commonly utilized tool is amyloid positron emission tomography (PET) which allows for the topographical visualization of amyloid-beta deposition in the brain. This type of scan can be done with or without [^{11}C] Pittsburgh compound-B ([^{11}C]PIB), which is a tracer compound that helps visualize amyloid-beta [65]. In addition, fluorodeoxyglucose [FDG]-PET and magnetic resonance imaging (MRI) are used to identify glucose metabolism difference and temporal atrophy, respectively. Both of which are known consequences of AD [61].

5.3 Risk Factors and Prevention

Alzheimer's Disease is a multifactorial disorder with various associated risk factors. Some of the most common are highlighted in this section below.

5.3.1 Age

Some risk factors are classified as nonmodifiable. The most considerable of which is age. Starting at the age of 65, the risk of AD doubles every 5 years. Additionally, the prevalence of AD rises from 10% at 65 years old to 40% after 85 years old [1]. Aging results in a decrease in brain weight, volume and neuronal density. Additionally, aging can lead to the development of amyloid plaques and granulovacuolar degeneration in the hippocampus which are two of the major hallmarks of AD. Aging is also associated with the destruction of the blood brain barrier (BBB) which can elicit an immune reaction, increasing reactive oxygen species (ROS), damage-associated molecular patterns (DAMPs) and inflammation [2].

5.3.2 Sex

Another nonmodifiable risk factor is sex. Females make up about two thirds of the patients with Alzheimer's [36]. It is currently debated whether there is truly a sex-based difference in AD prevalence. Some researchers suggest that the higher rate of females with AD is due to their longer life expectancy, as age is the strongest risk factor for Alzheimer's [37]. Other literature investigates the potential shared mechanism of AD and osteoporosis development, as females are at higher risk for developing both diseases compared to their male counterparts. Inflammation appears to be the common link between the two. Aging and bone disorders such as osteoporosis can lead to increased inflammation, promoting AD progression [48].

5.3.3 Inflammation

Recent studies have shown the significant role that peripheral inflammation plays in the progression of AD. Specifically, peripheral infection caused by herpes simplex virus type 1, cytomegalovirus, Epstein-Barr virus, human immunodeficiency virus, severe acute respiratory syndrome coronavirus 2, *Porphyromonas gingivalis*, *Helicobacter pylori*, and *Toxoplasma gondii* have all been implicated in their association with AD.

Disruption of blood brain barrier, neurotoxin buildup and dysregulated neurogenesis are all likely involved in the mechanism by which these pathogens can promote AD [47].

5.3.4 Gut-Brain Axis

Current literature shows the significant role the gut-brain axis (GBA) may play in the development of AD. The GBA is a bidirectional network between which the brain and the gut communicate via the enteric and central nervous system [3]. Patients with AD and MCI have been shown to have a less diverse gut microbiota than healthy controls [4]. Additionally, the presence or lack thereof certain microbes has been correlated with either elevated or lowered levels of AD biomarkers [3].

5.3.5 Diet

Lifestyle is another strong risk factor for AD development. Specifically, diet can either be protective against or associated with the development of mental disorders such as AD. Diets high in fats such as mono- and polyunsaturated fatty acids, fiber and protein have been shown to increase levels of beneficial microbes like *Prevotella*, *Bacteroidetes*, and *Lactobacillus*. Additionally, this diet is shown to decrease levels of harmful microbial organisms like Firmicutes, *Escherichia coli*, and *Ruminococcus*. Whereas a diet high in fat, sugar and processed foods are shown to have the opposite effect [3].

Recent studies also indicate the protective effects of foods high in polyphenols. The antioxidant properties of polyphenols help mitigate inflammation. Polyphenols also help increase the diversity and overall composition of the gut microbiota. Some foods rich in polyphenol include grapes, berries, olives, apples, nuts, seeds, pomegranates, beans, tomatoes, red wine, coffee, and tea [3].

There are a few dietary plans that have been most thoroughly studied and show neuroprotective properties to help delay the progression of AD. Those dietary patterns are:

1. Mediterranean diet
2. DASH diet (Dietary Approaches to Stop Hypertension)
3. MIND diet (Mediterranean-DASH Intervention for Neurodegenerative Delay)
4. Ketogenic diet

Overall, recent literature shows that diets high in vegetables, beans/legumes, nuts, seeds, and fish can mitigate inflammatory reactions to help prevent or slow the development of AD [5, 6].

5.3.6 Physical Activity

Current literature does suggest chronic physical activity can help reduce the risk and/or slow the development of AD. But there is not a consensus that has been reached on the type, timing or intensity of the activity that is most beneficial. Some studies specifically suggest that aerobic activities are the most beneficial. Aerobic activities can encompass running, walking, biking, swimming, and playing tennis. Possible mechanisms by which aerobic activity may slow AD progression include increase brain glucose metabolism, angiogenesis, release of neurotrophic factors, antioxidant capabilities, and cerebral vascular flow as well as decrease inflammatory activity [7, 8]. At the same time, other research suggests that anaerobic, strength training, balance and meditative exercise could also be associated with reduced AD development/progression [8].

5.3.7 Down Syndrome

Down Syndrome (DS) is a genetic disorder caused by the presence of three copies (trisomy) of chromosome 21. Amyloid precursor protein (APP) is spliced to create amyloid-beta protein. AD is characterized by an accumulation of amyloid-beta plaques. The APP gene is located on chromosome 21. It is thought that the overexpression of APP and thus amyloid-beta is what puts individuals with Down Syndrome at a higher risk of developing AD [9]. Studies suggest that 40–80% of individuals with Down Syndrome are diagnosed with AD by their 50s–60s and 100% of those with DS have AD pathology including amyloid plaques and NFTs [10].

5.3.8 Genetics

Alzheimer's Disease can be inherited as an autosomal dominant disorder through mutations in one of three genes [11]:

1. APP gene on chromosome 21
2. Presenilin 1 (PSEN1) on chromosome 14
3. Presenilin 2 (PSEN2) on chromosome 1

These single gene mutations only account for about 5% of AD cases and are usually associated with early-onset AD cases. APP gene mutations result in the overexpression of APP and thus amyloid-beta leading to amyloid beta plaque accumulation. Presenilin 1 and 2 gene mutations disrupt the proper development of gamma-secretase, one of the major proteolytic enzymes involved splicing APP to form amyloid-beta [1].

Additionally, apolipoprotein E (ApoE) has been linked to sporadic Alzheimer's Disease. ApoE is involved in the transport and metabolism of cholesterol and phospholipids. In Alzheimer's, ApoE has been implicated in the accumulation of amyloid-beta plaques [12]. In fact, approximately half of all AD cases involve the ApoE gene [13]. There are three isoforms of the gene: ApoE2, ApoE3 and ApoE4. ApoE2 has been shown to be protective against developing AD, whereas ApoE4 is indicative of an elevated risk of AD [13].

5.3.9 Traumatic Brain Injury

Traumatic Brain Injury (TBI) commonly results in the hyperphosphorylation of tau and subsequent release of this protein extracellular. This mechanism follows that of AD and describes a tauopathy. Some studies suggest that this extracellular release of hyperphosphorylated tau is initially a short-term protective mechanism after a TBI. In the event of recurrent TBIs or untreated TBIs, this mechanism is perpetually activated, leading to degenerative effects [14]. Other literature notes that TBI are correlated with an accelerated age of onset of AD [50].

5.3.10 Metabolic Disorders

Diabetes Mellitus (DM) is one of the most common metabolic disorders. DM is either the result of insulin insensitivity or lack of insulin secretion. While the exact mechanism is still debated, current literature shows that DM can lead to mild cognitive impairment and eventually AD [56]. Some researchers suggest that DM and hyperinsulinemia may increase the rate of amyloid-beta plaque and neurofibrillary tangle deposition [66]. Another mechanism proposed highlights how oxidative stress from DM can trigger inflammatory reactions, which may in turn progress AD development. Additionally, others focus on the mitochondrial dysfunction seen in both DM and AD which can lead to calcium dysregulation and improper splicing of APP. Of interest, APP and insulin receptors are both cleaved by the same proteolytic enzyme, beta-secretase [57, 66].

Obesity is another metabolic disorder implicated in development of AD. Studies have shown a link between increased visceral adipose tissue (VAT) and increased neurodegeneration and cognitive decline [60]. Increased VAT metabolism has also been associated with increased amyloid-beta deposition levels [59].

5.3.11 Cardiovascular Disease

Cardiovascular Disease (CVD) has also been associated with AD development. One finding reports that 10-year atherosclerotic cardiovascular disease risk scores were correlated to neurodegeneration but not amyloid or tau levels [67]. Additionally, some CVD risk factors including hypertension, high LDL cholesterol levels, and low HDL cholesterol levels are also risk factors for AD [68].

5.4 Future Directions

Alzheimer's Disease is a complex disorder that can impact and has been associated with many areas of health including cardiovascular disease, metabolic disease, gut function and obesity. Future research should continue to investigate this association with the goal of illuminating shared mechanisms and risk factors. This will help propel the development of more effective treatment options for patients with Alzheimer's Disease more effectively. Additionally, there have been breakthrough developments in biomarkers for AD. This is a research area of great importance and should be further funded. The ability to identify/diagnose individuals with AD based on biological markers instead of clinical symptoms alone is very powerful. This tool could allow individuals to start receiving treatment before clinical symptoms manifest which could drastically improve one's prognosis.

References

[1] Kumar, A., J. Sidhu, A. Goyal, J. W. Tsao and C. Doerr. 2023. Alzheimer disease. [updated 2022 jun 5]. StatPearls [Internet]. Treasure Island (FL): StatPearls Publishing.

[2] Nelson, V. K., N. K. Jha, M. V. Nuli, S. Gupta, S. Kanna, R. M. Gahtani et al. 2024. Unveiling the impact of Aging on BBB and Alzheimer's disease: Factors and Therapeutic implications. Ageing Research Reviews 98: 102224.

[3] Dissanayaka, D. S., V. Jayasena, S. R. Rainey-Smith, R. N. Martins and W. B. Fernando. 2024. The role of diet and gut microbiota in Alzheimer's Disease. Nutrients 16(3): 412.

[4] Xi, J., D. Ding, H. Zhu, R. Wang, F. Su, W. Wu et al. 2021. Disturbed microbial ecology in Alzheimer's disease: Evidence from the gut microbiota and fecal metabolome. BMC Microbiol. 21: 226.

[5] Oliphant, K. and E. Allen-Vercoe. 2019. Macronutrient metabolism by the human gut microbiome: Major fermentation by-products and their impact on host health. Microbiome 7: 1–15.

[6] Van den Brink, A. C., E. M. Brouwer-Brolsma, A. A. Berendsen and O. van de Rest. 2019. The Mediterranean, Dietary Approaches to Stop Hypertension (DASH), and Mediterranean-DASH Intervention for Neurodegenerative Delay (MIND) diets are associated with less cognitive decline and a lower risk of Alzheimer's disease—A review. Adv. Nutr. 10: 1040–1065.

[7] Abdullahi, A., T. W. L. Wong and S. S. M. Ng. 2024. Understanding the mechanisms of disease modifying effects of aerobic exercise in people with Alzheimer's disease. Ageing Research Reviews 102202.

[8] Paillard, T., H. Blain and P. L. Bernard. 2024. The impact of exercise on Alzheimer's disease progression. Expert Review of Neurotherapeutics 1–10.

[9] Sukreet, S., M. S. Rafii and R. A. Rissman. 2024. From understanding to action: Exploring molecular connections of Down syndrome to Alzheimer's disease for targeted therapeutic approach. Alzheimer's and Dementia: Diagnosis, Assessment and Disease Monitoring 16(2): e12580.

[10] Salehi, A., J. W. Ashford and E. J. Mufson. 2016. The link between Alzheimer's disease and down syndrome. A historical perspective. Curr Alzheimer Res. 13(1): 2–6.

[11] Wu, L., P. Rosa-Neto, G. Y. R. Hsiung, A. D. Sadovnick, M. Masellis, S. E. Black et al. 2012. Early-onset familial Alzheimer's disease (EOFAD). Canadian Journal of Neurological Sciences 39(4): 436–445.

[12] Lozupone, M. and F. Panza. 2024. Impact of apolipoprotein E isoforms on sporadic Alzheimer's disease: Beyond the role of amyloid beta. Neural Regeneration Research 19(1): 80–83.

[13] Stefanou, E. M. 2024. The role of ApoE in Alzheimer's Disease. Student International Biological and Biomedical Research 1(1).

[14] Martin, S. P. and B. A. Leeman-Markowski. 2024. Proposed mechanisms of tau: relationships to traumatic brain injury, Alzheimer's disease, and epilepsy. Frontiers in Neurology 14: 1287545.

[15] Brugnolo, A., B. Orso, N. Girtler, P. M. Ferraro, D. Arnaldi, P. Mattioli et al. 2024. Tracking the progression of Alzheimer's disease: Insights from metabolic patterns of SOMI stages. Cortex 171: 413–422.

[16] Serrano-Pozo, A., M. P. Frosch, E. Masliah and B. T. Hyman. 2011. Neuropathological alterations in Alzheimer disease. Cold Spring Harb Perspect Med. 1(1): a006189.

[17] Porsteinsson, A. P., R. S. Isaacson, S. Knox, M. N. Sabbagh and I. Rubino. 2021. Diagnosis of early Alzheimer's disease: clinical practice in 2021. The Journal of Prevention of Alzheimer's Disease 8(3): 371–386.

[18] Dubois, B., H. Hampel, H. H. Feldman, P. Scheltens, P. Aisen, S. Andrieu et al. 2016. Preclinical Alzheimer's disease: definition, natural history, and diagnostic criteria. Alzheimer's & Dementia 12(3): 292–323.

[19] Dubois, B. 2018. The emergence of a new conceptual framework for Alzheimer's disease. J. Alzheimers Dis. 62: 1059–66.

[20] Knopman, D. S., Parisi, J. E., Salviati, A., Floriach-Robert, M., Boeve, B. F., Ivnik, R. J. et al. 2003. Neuropathology of cognitively normal elderly. Journal of Neuropathology & Experimental Neurology 62(11): 1087–1095.

[21] Pais, M., L. Martinez, O. Ribeiro, J. Loureiro, R. Fernandez, L. Valiengo et al. 2020. Early diagnosis and treatment of Alzheimer's disease: new definitions and challenges. Brazilian Journal of Psychiatry 42(4): 431–441.
[22] Budelier, M. M. and R. J. Bateman. 2020. Biomarkers of Alzheimer disease. The Journal of Applied Laboratory Medicine 5(1): 194–208.
[23] Jack, C. R. Jr, M. S. Albert, D. S. Knopman, G. M. McKhann, R. A. Sperling, M. C. Carillo et al. 2011. Introduction to revised criteria for the diagnosis of Alzheimer's disease: national institute on aging and the Alzheimer association workgroups. Alzheimers Dement. 7: 257–62.
[24] Mian, M., J. Tahiri, R. Eldin, M. Altabaa, U. Sehar and P. H. Reddy. 2024. Overlooked cases of mild cognitive impairment: Implications to early Alzheimer's disease. Ageing Research Reviews 98: 102335.
[25] Vermunt, L., S. A. Sikkes, A. Van Den Hout, R. Handels, I. Bos, W. M. Van Der Flier et al. 2019. Duration of preclinical, prodromal, and dementia stages of Alzheimer's disease in relation to age, sex, and APOE genotype. Alzheimer's & Dementia 15(7): 888–898.
[26] Ye, B. S., H. J. Kim, Y. J. Kim, N. Y. Jung, J. S. Lee, J. Lee et al. 2018. Longitudinal outcomes of amyloid positive versus negative amnestic mild cognitive impairments: a three-year longitudinal study. Scientific Reports 8(1): 5557.
[27] Roberts, R. O., J. A. Aakre, W. K. Kremers, M. Vassilaki, D. S. Knopman, M. M. Mielke et al. 2018. Prevalence and outcomes of amyloid positivity among persons without dementia in a longitudinal, population-based setting. JAMA Neurology 75(8): 970–979.
[28] Marvi, F., Y. H. Chen and M. Sawan. 2024. Alzheimer's disease diagnosis in the preclinical stage: normal aging or dementia. IEEE Reviews in Biomedical Engineering, PP, 10.1109/RBME.2024.3376835. Advance Online Publication.
[29] Knopman, D. S., B. F. Boeve and R. C. Petersen. 2003, October. Essentials of the proper diagnoses of mild cognitive impairment, dementia, and major subtypes of dementia. In Mayo Clinic Proceedings 78(10): 1290–1308, Elsevier.
[30] Knopman, D. S. and R. C. Petersen. 2014, October. Mild cognitive impairment and mild dementia: a clinical perspective. In Mayo Clinic Proceedings 89: 1452–1459, Elsevier.
[31] Huang, S. S., W. F. Wang and Y. C. Liao. 2017. Severity and prevalence of behavioral and psychological symptoms among patients of different dementia stages in Taiwan. Archives of Clinical Psychiatry (São Paulo) 44: 89–93.
[32] Singh, H., C. Chopra, H. Singh, V. Malgotra, A. K. Wani, D. S. Dhanjal et al. 2024. Gut-brain axis and Alzheimer's disease: Therapeutic interventions and strategies. Journal of Functional Foods 112: 105915.
[33] Hippius, H. and G. Neundörfer. 2003. The discovery of Alzheimer's disease. Dialogues in Clinical Neuroscience 5(1): 101–108. https://doi.org/10.31887/DCNS.2003.5.1/hhippius.
[34] Chen, G. F., T. H. Xu, Y. Yan, Y. R. Zhou, Y. Jiang, K. Melcher et al. 2017. Amyloid beta: structure, biology and structure-based therapeutic development. Acta Pharmacologica Sinica 38(9): 1205–1235.
[35] Uddin, K. M. M., M. J. Alam, M. A. Uddin and S. Aryal. 2023. A novel approach utilizing machine learning for the early diagnosis of Alzheimer's disease. Biomedical Materials & Devices 1(2): 882–898.
[36] Alzheimer's Association: 2017. Alzheimer's disease facts and figures. Alzheimers Dement 13: 325–373.
[37] Mielke, M. M. 2018. Sex and gender differences in Alzheimer's disease dementia. The Psychiatric Times 35(11): 14.
[38] Corder, E. H., E. Ghebremedhin, M. G. Taylar, D. R. Thal, T. G. Ohm and H. Braak. 2004. The biphasic relationship b/w regional brain senile plaque and neurofibrillary Tangle distribution: modification by age, sex and APOE polymorphism. Ann N Y Acad Sci. 1019: 24–8.
[39] Sajjad, R., R. Arif, A. A. Shah, I. Manzoor and G. Mustafa. 2018. Pathogenesis of Alzheimer's disease: role of amyloid-beta and hyperphosphorylated tau protein. Indian J. Pharm. Sci. 80(4): 581–591.
[40] Hoozemans, J. J. M., R. Veerhuis, J. M. Rozemuller and P. Eikelenboom. 2006. Neuroinflamation and regeneration in the early stages of Alzheimer's disease pathology. Int. J. Dev. Neurosci. 24: 157–65.

[41] Andrade-Moraes, C. H., A. V. Oliveira-Pinto, E. Castro-Fonseca, C. G. da Silva, D. M. Guimarães, D. Szczupak et al. 2013. Cell number changes in Alzheimer's disease relate to dementia, not to plaques and tangles. Brain 136: 3738–3752.
[42] Serrano-Pozo, A., M. P. Frosch, E. Masliah and B. T. Hyman. 2011. Neuropathological alterations in Alzheimer disease. Cold Spring Harb Perspect Med 1: 123. https://doi.org/10.1101/cshperspect.a006189.
[43] Šimić, G., I. Kostović, B. Winblad and N. Bogdanović. 1997. Volume and number of neurons of the human hippocampal formation in normal aging and Alzheimer's disease. J. Comp Neurol 379: 482–494.
[44] Balusu, S. and B. De Strooper. 2024. The necroptosis cell death pathway drives neurodegeneration in Alzheimer's disease. Acta Neuropathologica 147(1): 96.
[45] Thal, D. R., K. Del Tredici, A. C. Ludolph, J. J. M. Hoozemans, A. J. Rozemuller, H. Braak et al. 2011. Stages of granulovacuolar degeneration: their relation to Alzheimer's disease and chronic stress response. Acta Neuropathol. 122: 577–589.
[46] Funk, K. E., R. E. Mrak and J. Kuret. 2011. Granulovacuolar degeneration (GVD) bodies of Alzheimer's disease (AD) resemble late-stage autophagic organelles. Neuropathology and Applied Neurobiology 37(3): 295–306.
[47] Li, Z., H. Wang and Y. Yin. 2024. Peripheral inflammation is a potential etiological factor in Alzheimer's disease. Reviews in the Neurosciences 35(1): 99–120.
[48] Karnik, S. J., T. J. Margetts, H. S. Wang, A. Movila, A. L. Oblak, J. C. Fehrenbacher et al. 2024. Mind the gap: unraveling the intricate dance between Alzheimer's disease and related dementias and bone health. Current Osteoporosis Reports 1–12.
[49] D'Souza, G. M., N. W. Churchill, D. X. Guan, M. A. Khoury, S. J. Graham, S. Kumar et al. 2024. Interaction of Alzheimer disease and traumatic brain injury on cortical thickness. Alzheimer Disease and Associated Disorders 38(1): 14–21.
[50] Rani, S., S. B. Dhar, A. Khajuria, D. Gupta, P. K. Jaiswal, N. Singla et al. 2023. Advanced overview of biomarkers and techniques for early diagnosis of alzheimer's disease. Cellular and Molecular Neurobiology 43(6): 2491–2523.
[51] Shafiee, N., V. Fonov, M. Dadar, R. N. Spreng and D. L. Collins. 2024. Degeneration in Nucleus basalis of Meynert signals earliest stage of Alzheimer's disease progression. Neurobiology of Aging 139: 54–63.
[52] Hampel, H., M. M. Mesulam, A. C. Cuello, M. R. Farlow, E. Giacobini, G. T. Grossberg et al. 2018. The cholinergic system in the pathophysiology and treatment of Alzheimer's disease. Brain 141(7): 1917–1933.
[53] Area-Gomez, E. and E. A. Schon. 2024. Towards a Unitary Hypothesis of Alzheimer's disease pathogenesis. Journal of Alzheimer's Disease (Preprint) 1–33.
[54] Barthélemy, N. R., G. Salvadó, S. E. Schindler, Y. He, S. Janelidze, L. E. Collij et al. 2024. Highly accurate blood test for Alzheimer's disease is similar or superior to clinical cerebrospinal fluid tests. Nature Medicine 30(4): 1085–1095.
[55] Jansen, W. J., R. Ossenkoppele, D. L. Knol, B. M. Tijms, P. Scheltens, F. R. Verhey et al. 2015. Amyloid Biomarker Study Group. Prevalence of cerebral amyloid pathology in persons without dementia: A meta-analysis. JAMA 313: 1924–1938.
[56] Zhang, Y., N. Q. Huang, F. Yan, H. Jin, S. Y. Zhou, J. S. Shi et al. 2018 Diabetes mellitus and Alzheimer's disease: GSK-3β as a potential link. Behav Brain Res. 339: 57–65.
[57] Bao, H., Y. Liu, M. Zhang, Z. Chen, W. Zhang, Y. Ge et al. 2021. Increased β-site APP cleaving enzyme 1-mediated insulin receptor cleavage in type 2 diabetes mellitus with cognitive impairment. Alzheimers Dement. 17: 1097–1108.
[58] Khanna, D., C. Peltzer, P. Kahar and M. S. Parmar. 2022. Body mass index (BMI): A screening tool analysis. Cureus 14(e22119).
[59] Zhou, S., L. Tu, W. Chen, G. Yan, H. Guo, X. Wang et al. 2024. Alzheimer's disease, a metabolic disorder: Clinical advances and basic model studies. Experimental and Therapeutic Medicine 27(2): 1–10.
[60] Prem Kumar, A., N. Singh, D. Nair and A. Justin. 2022. Neuronal PET tracers for Alzheimer's disease. Biochem Biophys Res. Commun. 587: 58–62.

[61] Arslan, B., H. Zetterberg and N. Ashton. 2024. Blood-based biomarkers in Alzheimer's disease – moving towards a new era of diagnostics. Clinical Chemistry and Laboratory Medicine (CCLM) 62(6): 1063–1069. https://doi.org/10.1515/cclm-2023-1434.
[62] Gonzalez-Ortiz, F., M. Turton, P. R. Kac, D. Smirnov, E. Premi, R. Ghidoni et al. 2023. Brain-derived tau: a novel blood-based biomarker for Alzheimer's disease-type neurodegeneration. Brain 146: 1152–65. https://doi.org/10.1093/brain/awac407.
[63] Benedet, A. L., M. Milà-Alomà, A. Vrillon, N. J. Ashton, T. A. Pascoal, F. Lussier et al. 2021. Differences between plasma and cerebrospinal fluid glial fibrillary acidic protein levels across the Alzheimer disease continuum. JAMA Neurol 78: 1471–83.
[64] Pereira, J. B., S. Janelidze, R. Smith, N. Mattsson-Carlgren, S. Palmqvist, C. E. Teunissen et al. 2021. Plasma GFAP is an early marker of amyloid-β but not tau pathology in Alzheimer's disease. Brain 144: 3505–16. https://doi.org/10.1093/brain/awab223.
[65] Nordberg, A., S. F. Carter, J. Rinne, A. Drzezga, D. J. Brooks and R. Vandenberghe. 2013. A European multicentre PET study of fibrillar amyloid in Alzheimer's disease. European Journal of Nuclear Medicine and Molecular Imaging 40: 104–114.
[66] Sims-Robinson, C., B. Kim, A. Rosko and E. L. Feldman. 2010. How does diabetes accelerate Alzheimer disease pathology? Nature Reviews Neurology 6(10): 551–559.
[67] Saeed, A., Y. Chang, J. Swanson, M. Vu, M. Mapstone, V. L. Villemagne et al. 2024. Longitudinal Association of Mid-Life Ten Year Cardiovascular Disease Risk Score with Brain Biomarkers of Alzheimer's Disease, Neurodegeneration and White Matter Hyper Intensities in Cognitively Unimpaired Older Adults: Heart SCORE Brain Study. medRxiv, 2024-01.
[68] Stampfer, M. J. 2006. Cardiovascular disease and Alzheimer's disease: common links. Journal of Internal Medicine 260(3): 211–223.
[69] Isik, A. T. 2010. Late onset Alzheimer's disease in older people. Clinical Interventions in Aging 5: 307–311. https://doi.org/10.2147/CIA.S11718.
[70] Gustavsson, A., N. Norton, T. Fast, L. Frölich, J. Georges, D. Holzapfel et al. 2023. Global estimates on the number of persons across the Alzheimer's disease continuum. Alzheimer's and Dementia 19(2): 658–670.
[71] Hebert, L. E., J. Weuve, P. A. Scherr and D. A. Evans. 2013. Alzheimer disease in the United States (2010–2050) estimated using the 2010 census. Neurology 80(19): 1778–1783. doi:10.1212/WNL.0b013e31828726f5.

CHAPTER 6

Autism Spectrum Disorder

Sophia Marra,[1,*] *Percin Karakol*[2] and *Emre Tokgoz*[3]

6.1 Introduction

Autism Spectrum Disorder (ASD) is a multifactorial neurodevelopmental disorder. ASD is characterized by social communication difficulties in addition to repetitive and restrictive behaviors [1]. Usually, ASD symptoms are noticed early in development and are present throughout life [28]. The term "autism" was first coined by Dr. Eugen Bleuler to describe "infantile wishes" and hallucinations as symptoms of schizophrenia [59]. Then, through the work of many child psychologists in the 1960s, the term was redefined to describe the exact opposite: a lack of fantasizing [60].

Of note, Dr. Leo Kanner and Micheal Rutter are credited with groundbreaking work to understand and define autism. Dr. Kanner in 1943 described 11 individuals who he described as having "autistic disturbance of affect contact" [61]. Specifically, he noted that these children had difficulties in the areas of relationship development, speech development and non-communicative speech usage [11]. He coined the term "early infantile autism" [61]. Dr. Rutter outlined four hallmarks of infantile autism as: (1) speech deficits (2) lack of interest in social interaction (3) repetitive and restrictive motor behaviors (4) early onset before 30 months old [62].

Dr. Hans Asperger published work in 1944 describing what he called "autistic psychopathy" after observing four boys aged 6–11 years old [63]. Asperger appeared to edit the defining characteristics of the syndrome he observed throughout his career. By 1979, he highlighted characteristics of high intelligence, logic and abstract

[1] Frank H. Netter School of Medicine, Quinnipiac University, North Haven, CT, 06473, USA.
[2] Başakşehir Cam and Sakura City Hospital Plastic Reconstructive and Aesthetic Surgery, Istanbul, Turkiye.
[3] School of Engineering and Technology, Farmingdale State College, New York, 11735 USA.
* Corresponding author: Sophia.Marra@quinnipiac.edu

DOI: 10.1201/9781003646044-6

thinking [64]. As of 2013, the DSM-5 no longer includes Asperger's as a separate diagnosis [28]. Now, autism is seen as a spectrum of characteristics and behaviors that will be described in detail below.

6.2 Clinical Presentation

Autism Spectrum Disorder (ASD) is neurodevelopmental disorder characterized by social communication challenges as well as repetitive behaviors and interests [1]. The global prevalence of ASD is 1%, with increased percentages seen in higher-income countries [2]. For example, in the United States, current research suggests that 1 in every 36 (2.8%) 8-year-olds has ASD [4]. The ratio of females with ASD compared to males is 1 to 4.2 [2]. Interestingly, over the past 20 years, the proportion of females diagnosed with ASD is growing more rapidly than males [5]. This may reflect an increase in information available on how the disorder may manifest in females compared to males.

Currently there are no verified biomarkers for ASD, so diagnoses are made based on behavioral symptoms. ASD symptoms can be identified as early as 18 to 24 months old [2]. Current literature suggests that ASD symptoms are not present at birth, but rather there is regression over time [3]. This can include new deficiencies in language, social connection, or skills in previously proficient areas [6]. It is common for adults to notice certain potential symptoms of ASD in infants around the age of 12 mo. Some common early signs seen in infants by 12 months of age include [7, 8]:

- **Not responding to their name**
- **Lack of gestures-** This can include a lack of pointing at desired objected or to get one's attention. Additionally, infants at this age also usually gesture to be picked in, wave and shake their head to motion "no".
- **Limited play with toys**
- **Not babbling-** This usually includes vowel and consonant combinations like "ba", "ma", "da" and "gee"
- **Not following an adult's pointed finger**
- **Repetitive actions-** This can include repetitive flapping of one's arms or repetitively playing with a toy inappropriately
- **Poor eye contact**
- **Lack of interaction with others**

Current literature suggests that intervention as early as 6 months of age can lead to improvement in ASD symptomology in certain areas [9]. This has given credence to the identification of early signs of ASD as listed above. There is limited information regarding the progression and course of ASD throughout one's life [12, 13].

6.3 Diagnosis

6.3.1 Current Diagnostic Criteria

According to American Psychiatric Association's *Diagnostic and Statistical Manual,* Fifth Edition (DSM-5), to be diagnosed with ASD, a child must have "persistent deficits in each of three areas of social communication" [10]:

1. *Social-emotional reciprocity*- Some examples may include difficulty conversing or lack of expressing interests and emotions.
2. *Nonverbal communicative behaviors*- This may manifest as an inability to understand and/or use of facial expression, eye contact or body language.
3. *Developing, Maintaining and understanding relationships*- Example can include difficulty adjusting behavior for individual social situations or a lack of interest in others.

Additionally, the child must exhibit "at least two of the four types of restricted, repetitive behavior:

1. Stereotyped or repetitive motor movements, use of objects, or speech
2. Insistence of sameness, inflexible adherence to routines or ritualized patterns of verbal or nonverbal behavior
3. Highly restricted, fixated interests that are abnormal in intensity or focus
4. Hyper- or hypo-reactivity to sensory input or unusual interest in sensory aspects of the environment" [10]

ASD is a disorder that is viewed on a spectrum, so the DSM-5 also stipulates identification of the current severity of communication deficits and restrictive patterns. There are three levels possible [10]:

1. Level One- Requires support
2. Level Two- Requires significant care
3. Level Three- Requires very significant care

The 5th edition of the DSM was the first time that autism was defined as a spectrum. The spectrum now includes the previously separate Asperger's disorder, pervasive developmental disorder not otherwise specified (PDDNOS) and childhood disintegrative disorder [11].

6.3.2 ASD Presentation in Females

As previously mentioned, there is a 4:1 discrepancy in the number of males diagnosed with ASD compared to females. Current literature proposes two possible explanations for this finding. One plausible reason for this difference is that females possess some sort of factor that protects them from the development of ASD. Others postulate that the clinical presentation and phenotype of ASD in females may differ from the currently accepted criteria.

6.3.2.1 Female Protective Effect (FPE) Theory

The female protective effect (FPE) theory describes the postulation that females have a lower incidence of autism than males because of a protective factor or collection of such factors [14]. This idea theorizes that females require a greater level of environmental and/or genetic risk factors present to have the same severity of ASD symptoms as male counterparts [15]. This theory is grounded in current research that has shown that females with autism appear to have "relatively more spontaneous, non-inherited mutations associated with autism" as compared to males [16, 17]. Importantly, the females and males in these studies had the same degree of ASD characteristics.

Current literature in support of this theory has shown that the first relatives of females with autism are more likely to have autism themselves as compared to those relatives of males with autism. The FPE theory suggests that females require a greater genetic load to develop ASD characteristics. Recent research has also shown the strong inheritable genetic component of ASD development [18]. The first relatives of females with autism should thus also have a greater genetic load for autism compared to those of males with autism. Some studies have indeed found this to be the case while others have found the opposite [19, 20].

One issue with this theory is that this "protective factor" or a collection of factors has not yet been identified. Some researchers suggest that androgen levels may correlate with autism characteristics [14]. This idea is grounded in the extreme male brain theory (EMB) theory which postulates that there is an association with androgens and some of the characteristics of autism [21]. The EMB theory also suggests that conversely those with lower levels of androgens (females) also have lower levels of autism characteristics [22]. Thus, lower levels of androgens may be protective against the development of autism. In fact, some current research suggests that higher levels of androgens in females may be correlated with greater levels of autism characteristics [23]. Conversely, other studies have found there to be limited or no correlation between the two factors [24].

Current literature is also investigating the possibility of a protective gene that may be on the X chromosome. Specifically, some research suggests that the lower prevalence of autism in females may be explained by a protective gene inherited on the paternal X chromosome [26]. So far, one such X chromosome gene has been identified. *TBL1X* is known to be involved in the Wnt signaling pathway. It has also been identified to potentially play a role in autism. Specifically, deletions in the Xp22.2 to Xp22.3 region containing *TBL1X* have been implicated in the development of disorders including ASD and intellectual disabilities [25].

On the other hand, other research suggests that the sex-based differences are not due to any protective factors in females at all. Rather this discrepancy is due to a difference in presentation of autism in females. Thus, these females are not diagnosed because current diagnostic criteria is based on primarily male cohort and presentation [27].

6.3.2.2 Female Autism Phenotype Theory

The female autism phenotype theory suggests that the discrepancy of ASD diagnosis in females versus males is due to a difference in autism presentation in females. Research in this area proposes that females may exhibit the diagnostic characteristics of autism like social communication deficits and repetitive behavior but in different ways than traditional examples [14]. Also, current literature suggests that females with autism may show additional characteristics not currently included in the diagnostic criteria. The most common characteristics include:

- **Social Relationships.** Current research suggests that females with autism may have greater motivation to make friends compared to males with autism [29]. Furthermore, studies suggest that females with autism have the same social motivation as females without autism [30]. This is very different than the hallmark presentation of ASD of a disinterest and lack of motivation to make relationships with peers [28]. However, the presentation of ASD may differ as females with autism seem to have difficulty maintaining long-term relationships even though they have the desire to continue [30]. Additionally, females with autism may also have more trouble handling conflict in relationships as compared to males with autism and females without autism [31].
- **Repetitive and/or Restrictive Behavior.** Recent research has been done to investigate sex-based differences in repetitive and/or restrictive behaviors- another hallmark of ASD. Some studies suggest that females with autism show lower levels of these behaviors [32]. At the same time, others postulate that females do in fact show repetitive and/or restrictive behavior, but they go undocumented and undiagnosed because they differ from those commonly seen in males with autism [33]. Specifically, researchers have found that males with autism tend to show restrictive and/or repetitive behaviors in areas related to vehicles, computers or physics [34, 35]. Whereas females with autism more commonly demonstrate these behaviors in the areas of animals, fictional characters or psychology [34, 35]. Because these areas of interest are not thought of as "typical" as it related to the current understanding and diagnosis of ASD, females might be less likely to be diagnosed [36]. Additionally, these content of interest in females may be perceived as more socially acceptable or age-appropriate and thus may not be identified as abnormal by parents or clinicians despite that fact that the intensity of the interest in representative of that seen in ASD [14].
- **Internalizing Problems.** Internalizing problems refers to the inward-rather than outward- "expression of emotional difficulties, including anxiety, depression, self-harming and eating disorders" [14]. This symptom falls under the category of additional characteristics seen in females with autism rather than category of characteristics expressed differently in females (like the first two described above). Current studies suggest that females with autism are at a significantly higher risk of developing internalizing disorders than males [37, 38]. In comparison, males with autism are a significantly higher risk of developing externalizing disorders like behavioral problems and inattentiveness [30]. As a

result, females may not be diagnosed as frequently as their male counterparts for one of two reasons. First, if internalization is seen as an "abnormal" characteristic of autism it may be missed by family, teachers and clinicians as a symptom of ASD. Instead, females may only be diagnosed with their comorbid internalizing disorder as their ASD goes undiagnosed. Additionally, because internalization behaviors are commonly less disruptive than externalization behaviors, there may be less of a drive to identify and refer females who internalize emotions compared to those with more disruptive, externalizing behaviors [14].

6.3.2.3 Camouflaging

One important consideration in the conversation of autism presentation in women is the concept of camouflaging. Camouflaging refers to the implementation, whether consciously or not, of strategies to minimize or hide autism characteristics that stand out in social settings [41]. The two key aspects of camouflaging include masking (hiding of autism characteristics) and compensation (implementing strategies to make for social and communication difficulties) [39]. The idea of compensation was developed concurrently as camouflaging. It specifically refers to the use of executive reasoning, instead of social intellect, to behave in line with social norms and mask autism-related deficits [40]. Studies suggest that these socially "correct" behaviors may be recognized and learned from peers [14]. Some proposed examples of camouflaging include [14]:

1. Mimicking facial expressions appropriate for the social situation
2. Suppression of physical behaviors such as hand flapping
3. Reciting scripted responses in social interactions
4. Forcing oneself to maintain eye contact
5. Stop oneself from taking about interests, especially as it relates to repetitive and/or restrictive behavior

It is especially important to note that many individuals find camouflaging stressful and exhausting, whether this is a conscious effort or not [40]. The implementation of these camouflaging behaviors may perpetuate a feeling that individuals with autism must adapt and hide elements of themselves which can lower one's self esteem [41].

While this type of behavior can be seen in both sexes, females have been reported to employ this tactic more often. There are varying explanations presented for this sex-based difference in behavior. Some report that females with autism may be under more pressure to fit in socially compared to males with autism based on societal and social expectations placed on females [42]. Others propose that differences in "cognitive profiles" may give females with autism a greater ability to engage in camouflaging behaviors [43]. Still others suggest that all individuals, including those that do not identify within the binary system of female vs male, may use camouflaging strategies [44]. There is a need for continued research to investigate the role of gender and sex in camouflaging behavior. Additionally, this research must continue to include the experience and perspective of those that do not fall within the gender binary.

Current literature suggests that the use of camouflaging in social settings may be another reason females are less likely to be diagnosed compared to males. The individual social setting is particularly important when considering the implementation of camouflaging tactics. For example, when in a familiar, low-social pressure environment, females with autism may feel comfortable to "drop the mask" and not feel the need to adapt. Conversely, when in higher pressure social situations, some may find themselves compensating or adapting to fit in with social norms. Thus, it is important to consider the varying expression of autism characteristics in different environments. Camouflaging strategies may differ at home, with friends or family, in school or in clinical settings which can make diagnosis more challenging [14, 32, 37, 45]. This in turn may be a key factor at play when considering the sex-based discrepancy in ASD diagnosis.

Recent research initiatives have included the implementation of clinical measures to identify camouflaging behaviors. Currently, the two main ways that camouflaging is measured are listed below:

- **Discrepancy Method.** This model of measuring camouflaging behaviors involves the identification of both the internal and external autism presentation. The internal presentation refers to one's true, unedited autism characteristics without any adaptation or suppression. Conversely, external presentation is how one shows themselves to others in social situations. This external view differs from the internal via the implementation of camouflaging strategies. Thus, it is the *discrepancy* between these two behavioral presentations that is measured.

  Current research using the discrepancy method suggests that females with autism have a larger discrepancy between self-report autism traits and autism behaviors reported by an observer [46]. But one of the weaknesses of this approach is that self-reporting of internal autism characteristics is not a scientifically reliable method of identifying true, unbiased internal behaviors. Additionally, this approach only measures successful camouflaging strategies instead rather than including all attempts, even those that were not successful [14].
- **Observational Method.** In comparison, the observational method strives to identify camouflaging behaviors themselves, rather than discrepancies in autistic presentation. The identification of such behaviors is done by the individual themselves and/or others close to them. This approach also is not dependent on the success of camouflaging strategies and instead simply indicates the use of such behaviors, unlike the discrepancy method [14]. One such measure that has been validated is the Camouflaging Autistic Traits Questionnaire (CAT-Q). This questionnaire has only been validated for an adult population. One strength is that the CAT-Q can be used females and males with and without autism allowing analysis across groups [47].

6.4 Risk Factors of ASD

Autism Spectrum Disorder is a heterogeneous, complex disability that has been linked to both genetic and non-genetic risk factors. However, the exact etiology of ASD is not fully understood. The multifactorial components are thought to include age of parents, genetics, comorbidities, dysfunctional immune system, adverse environment, heavy metal toxicity, mitochondrial dysfunction, and prenatal, perinatal and postnatal factors [12].

Because ASD is a neurodevelopmental disability, there is a great deal of research that has investigated what time points in the development of the brain are most critical when considering ASD development. Some research suggests that since many autism diagnoses are given before the age of 2 years old, this critical window must fall between prenatal and very early postnatal periods [49]. Recent findings have pointed to first trimester as an important timeframe where there is increased developmental vulnerability [48]. Studies have also shown that increased parental age and decreased intervals between pregnancies are associated with the development of autism [51, 52].

Current literature has shown an association between prenatal maternal stress and immune dysregulation and the development of ASD [50]. But importantly, these exposures do not cause ASD, which points to the likelihood of many other factors. Specifically, this finding indicates a potentially important gene x environment relationship. This means that some populations exposed to psychological stressors may be more vulnerable to having children with autism compared to others. As mentioned above, the maternal immune system is also associated with ASD development. When the maternal immune system is disrupted, the release of cytokines, chemokines and antibodies has been shown to have negative effects on the developing fetus, impacting neurodevelopment and behavior [50].

In addition to maternal stress and immune dysregulation, there are other maternal prenatal exposures associated with ASD. The drugs that have been identified in current research to increase the risk of ASD include [50, 53, 54, 55, 56, 57, 58]:

1. *Pollutants*: Air pollutants in particular have been linked to ASD development. Specifically, air pollutants have been shown to interact with the tyrosine kinase MET receptor gene.
2. *Valproic acid*
3. *β2-adrenergic agonists*: These drugs are commonly given to stop premature labor.
4. *Pesticides*
5. *Endocrine disrupting chemicals (EDCs)*: Studies have specifically linked polychlorinated biphenyls and phthalates with autism.
6. *Dietary factors*: Specifically, a lack of folate supplementation has been shown to be associated with ASD development.

As noted above, ASD is a multifactorial complex disorder that is likely a result of many genetic and non-genetic factors.

6.5 Future Directions

As the prevalence of ASD continues to grow, it is important that we continue to understand this multifaceted disorder. Current literature must continue to investigate the significant discrepancy in females versus males with autism. As is the case with many areas of research, much of what we understand about disorders is based on a male cohort, so we must emphasize the importance of female representation in research. Furthermore, current data is commonly reported within the gender binary. It is important for us to understand how autism presents and affects individuals of all sex and gender, including those who do not identify within the binary. The literature suggests that the autism characteristics in females tend to be less disruptive and externally obvious, thus they are less likely to be referred for diagnosis. This is why it is important to be able to identify all presentations, even those not classically defined.

Finally, future research should continue to investigate the etiology of the disorder. ASD is a heterogenous disorder which suggests the presents of various pathways and factors involved. In order to best care for patients with ASD, we must understand the underlying mechanisms.

References

[1] Cook, J., L. Hull, L. Crane and W. Mandy. 2021. Camouflaging in autism: A systematic review. Clinical Psychology Review 89: 102080.

[2] Zeidan, J., E. Fombonne, J. Scorah, A. Ibrahim, M. S. Durkin, S. Saxena et al. 2022. Global prevalence of autism: A systematic review update. Autism Research 15(5): 778–790.

[3] Ozonoff, S., A. M. Iosif, F. Baguio, I. C. Cook, M. M. Hill, T. Hutman et al. 2010. A prospective study of the emergence of early behavioral signs of autism. Journal of the American Academy of Child and Adolescent Psychiatry 49(3): 256–266.

[4] Maenner, M. J. 2023. Prevalence and characteristics of autism spectrum disorder among children aged 8 years—Autism and Developmental Disabilities Monitoring Network, 11 sites, United States, 2020. MMWR. Surveillance Summaries, 72.

[5] Harrop, C., B. Tomaszewski, O. Putnam, C. Klein, E. Lamarche and L. Klinger. 2024. Are the diagnostic rates of autistic females increasing? An examination of state-wide trends. Journal of Child Psychology and Psychiatry.

[6] Zwaigenbaum, L., S. Bryson, C. Lord, S. Rogers, A. Carter, L. Carver et al. 2009. Clinical assessment and management of toddlers with suspected autism spectrum disorder: insights from studies of high-risk infants. Pediatrics 123(5): 1383–1391.

[7] Watson, L. R. and E. R. Crais. 2013. Translating between research and practice in serving infants at risk for ASD. Perspectives on Language Learning and Education 20(1): 4–13.

[8] Zwaigenbaum, L., S. Bryson, C. Lord, S. Rogers, A. Carter, L. Carver et al. 2009. Clinical assessment and management of toddlers with suspected autism spectrum disorder: insights from studies of high-risk infants. Pediatrics 123(5): 1383–1391.

[9] Bradshaw, J., A. M. Steiner, G. Gengoux and L. K. Koegel. 2015. Feasibility and effectiveness of very early intervention for infants at-risk for autism spectrum disorder: A systematic review. Journal of Autism and Developmental Disorders 45: 778–794.

[10] Centers for Disease Control and Prevention. (n.d.). Clinical testing and diagnosis for autism spectrum disorder. Centers for Disease Control and Prevention. https://www.cdc.gov/autism/hcp/diagnosis/index.html.

[11] Tsai, L. Y. and M. Ghaziuddin. 2014. DSM-5 ASD moves forward into the past. Journal of Autism and Developmental Disorders 44: 321–330.

[12] Joon, P., A. Kumar and M. Parle. 2021. What is autism?. Pharmacological Reports 73(5): 1255–1264.
[13] Siebes, R., J. W. Muntjewerff and W. Staal. 2018. Differences of symptom distribution across adult age in high functioning individuals on the autism spectrum using subscales of the autism spectrum quotient. J. Autism Dev. Disord. 48(11): 3939–44.
[14] Hull, L., K. V. Petrides and W. Mandy. 2020. The female autism phenotype and camouflaging: A narrative review. Review Journal of Autism and Developmental Disorders 7: 306–317.
[15] Robinson, E. B., P. Lichtenstein, H. Anckarsäter, F. Happé and A. Ronald. 2013. Examining and interpreting the female protective effect against autistic behavior. Proceedings of the National Academy of Sciences of the United States of America 110(13): 5258–5262. https://doi.org/10.1073/pnas.1211070110.
[16] Gilman, S. R., I. Iossifov, D. Levy, M. Ronemus, M. Wigler and D. Vitkup. 2011. Rare *De novo* variants associated with autism implicate a large functional network of genes involved in formation and function of synapses. Neuron 70(5): 898–907.
[17] Levy, D., M. Ronemus, B. Yamrom, Y. Lee, A. Leotta, J. Kendall et al. 2011. Rare *De novo* and transmitted copy-number variation in autistic Spectrum disorders. Neuron 70(5): 886–897. https://doi.org/10.1016/j.neuron.2011.05.015.
[18] Tick, B., P. Bolton, F. Happé, M. Rutter and F. Rijsdijk. 2016. Heritability of autism spectrum disorders: a meta-analysis of twin studies. Journal of Child Psychology and Psychiatry 57(5): 585–595.
[19] Desachy, G., L. A. Croen, A. R. Torres, M. Kharrazi, G. N. Delorenze, G. C. Windham et al. 2015. Increased female autosomal burden of rare copy number variants in human populations and in autism families. Molecular Psychiatry 20(2): 170–175. https://doi.org/10.1038/mp.2014.179.
[20] Ozonoff, S., G. Young, A. Carter, D. Messinger, N. Yirmiya, L. Zwaigenbaum et al. 2011. Recurrence risk for autism spectrum disorders: A baby siblings research consortium study. Pediatrics 128(3): 488–495. https://doi.org/10.1542/peds.2010-2825.
[21] Baron-Cohen, S. 2002. The extreme male brain theory of autism. Trends in Cognitive Sciences 6(6): 248–254. https://doi.org/10.1016/S1364-6613(02)01904-6.
[22] Baron-Cohen, S., B. Auyeung, B. Nørgaard-Pedersen, D. M. Hougaard, M. W. Abdallah, L. Melgaard et al. 2015. Elevated fetal steroidogenic activity in autism. Molecular Psychiatry 20(3): 369–376. https://doi.org/10.1038/mp.2014.48.
[23] Schwarz, E., P. C. Guest, H. Rahmoune, L. Wang, Y. Levin, E. Ingudomnukul et al. 2011. Sex-specific serum biomarker patterns in adults with Asperger's syndrome. Molecular Psychiatry 16: 1213–1220. https://doi.org/10.1038/mp.2010.102.
[24] Guyatt, A. L., J. Heron, B. L. C. Knight, J. Golding and D. Rai. 2015. Digit ratio and autism spectrum disorders in the Avon longitudinal study of parents and children: A birth cohort study. BMJ Open 5(8): e007433. https://doi.org/10.1136/bmjopen-2014-007433.
[25] Chung, R. H., D. Ma, K. Wang, D. J. Hedges, J. M. Jaworski, J. R. Gilbert et al. 2011. An X chromosome-wide association study in autism families identifies TBL1X as a novel autism spectrum disorder candidate gene in males. Molecular Autism 2: 1–10.
[26] Skuse, D. H. 2000. Imprinting, the X-chromosome, and the male brain: Explaining sex differences in the liability to autism. Pediatric Research 47: 9.
[27] Kirkovski, M., P. G. Enticott and P. B. Fitzgerald. 2013. A review of the role of female gender in autism spectrum disorders. Journal of Autism and Developmental Disorders 43(11): 2584–2603. https://doi.org/10.1007/s10803-013-1811-1.
[28] American Psychiatric Association. 2013. DSM 5. American Psychiatric Association.
[29] Head, A. M., J. A. McGillivray and M. A. Stokes. 2014. Gender differences in emotionality and sociability in children with autism spectrum disorders. Molecular Autism 5(19). https://doi.org/10.1186/2040-2392-5-19.
[30] Hiller, R. M., R. L. Young and N. Weber. 2014. Sex differences in autism spectrum disorder based on DSM-5 criteria: Evidence from clinician and teacher reporting. Journal of Abnormal Child Psychology 42(8): 1381–1393. https://doi.org/10.1007/s10802-014-9881-x.
[31] Sedgewick, F., V. Hill and E. Pellicano. 2019. 'It's different for girls': Gender differences in the friendships and conflict of autistic and neurotypical adolescents. Autism 23(5): 1119–1132. https://doi.org/10.1177/1362361318794930.

[32] Lai, M.-C., M. V. Lombardo, B. Auyeung, B. Chakrabarti and S. Baron-Cohen. 2015. Sex/gender differences and autism: Setting the scene for future research. Journal of the American Academy of Child and Adolescent Psychiatry 54(1): 11–24.
[33] Antezana, L., R. S. Factor, E. E. Condy, M. V. Strege, A. Scarpa and J. A. Richey. 2019. Gender differences in restricted and repetitive behaviors and interests in youth with autism. Autism Research 12(2): 274–283. https://doi.org/10.1002/aur.2049.
[34] Nowell, S. W., D. R. Jones and C. Harrop. 2019. Circumscribed interests in autism: Are there sex differences? Advances in Autism, AIA-09-2018-0032. https://doi.org/10.1108/AIA-09-2018-0032.
[35] Grove, R., R. A. Hoekstra, M. Wierda and S. Begeer. 2018. Special interests and subjective wellbeing in autistic adults. Autism Research 11(5): 766–775. https://doi.org/10.1002/aur.1931.
[36] Sutherland, R., A. Hodge, S. Bruck, D. Costley and H. Klieve. 2017. Parent-reported differences between school-aged girls and boys on the autism spectrum. Autism 21(6): 785–794. https://doi.org/10.1177/1362361316668653.
[37] Mandy, W., L. Pellicano, B. St Pourcain, D. Skuse and J. Heron. 2018. The development of autistic social traits across childhood and adolescence in males and females. Journal of Child Psychology and Psychiatry 59(11): 1143–1151. https://doi.org/10.1111/jcpp.12913.
[38] Oswald, T. M., M. A. Winter-Messiers, B. Gibson, A. M. Schmidt, C. M. Herr and M. Solomon. 2016. Sex differences in internalizing problems during adolescence in autism Spectrum disorder. Journal of Autism and Developmental Disorders 46(2): 624–636. https://doi.org/10.1007/s10803-015-2608-1.
[39] Fombonne, E. 2020. Camouflage and autism. Journal of Child Psychology and Psychiatry 61(7): 735–738.
[40] Cook, J., L. Hull, L. Crane and W. Mandy. 2021. Camouflaging in autism: A systematic review. Clinical Psychology Review 89: 102080.
[41] Hull, L., K. V. Petrides, C. Allison, P. Smith, S. Baron-Cohen, M. Lai et al. 2017a. "Putting on my best normal": Social camouflaging in adults with autism Spectrum conditions. Journal of Autism and Developmental Disorders 47(8): 2519–2534.
[42] Pearson, A. and K. Rose. 2021. A conceptual analysis of autistic masking: Understanding the narrative of stigma and the illusion of choice. Autism in Adulthood 3(1): 52–60.
[43] Lehnhardt, F. G., C. M. Falter, A. Gawronski, K. Pfeiffer, R. Tepest, J. Franklin et al. 2016. Sex-related cognitive profile in autism spectrum disorders diagnosed late in life: implications for the female autistic phenotype. Journal of Autism and Developmental Disorders 46: 139–154.
[44] Cook, J., L. Hull, L. Crane and W. Mandy. 2021. Camouflaging in autism: A systematic review. Clinical Psychology Review 89: 102080.
[45] Lai, M.-C., M. V. Lombardo, G. Pasco, A. N. V. Ruigrok, S. J. Wheelwright, S. A. Sadek et al. 2011. A behavioral comparison of male and female adults with high functioning autism spectrum conditions. PLoS One 6(6): e20835. https://doi.org/10.1371/journal.pone.0020835.
[46] Lai, M.-C., M. V. Lombardo, A. N. V. Ruigrok, B. Chakrabarti, B. Auyeung, P. Szatmari et al. 2017. Quantifying and exploring camouflaging in men and women with autism. Autism 21(6): 690–702. https://doi.org/10.1177/1362361316671012.
[47] Hull, L., W. Mandy, M. C. Lai, S. Baron-Cohen, C. Allison, P. Smith et al. 2018. Development and validation of the camouflaging autistic traits questionnaire (CAT-Q). Journal of Autism and Developmental Disorders 49(3): 819–833. https://doi.org/10.1007/s10803-018-3792-6.
[48] Johansson, M., E. Wentz, E. Fernell, K. Strömland, M. T. Miller and C. Gillberg. 2001. Autistic spectrum disorders in Möbius sequence: A comprehensive study of 25 individuals. Dev Med Child Neurol. 43: 338–45.
[49] Dietert, R. R., J. M. Dietert and J. C. DeWitt. 2011. Environmental risk factors for autism. Emerging Health Threats Journal 4(1): 7111.
[50] Beversdorf, D. Q., H. E. Stevens and K. L. Jones. 2018. Prenatal stress, maternal immune dysregulation, and their association with autism spectrum disorders. Current Psychiatry Reports 20: 1–12.
[51] D'Onofrio, B. M., M. E. Rickert, E. Frans, R. Kuja-Halkola, C. Almqvist, A. Sjölander et al. 2014. Paternal age at childbearing and offspring psychiatric and academic morbidity. JAMA Psychiatry 71: 432–8.

[52] Cheslack-Postava, K., A. Suominen, E. Jokiranta, V. Lehti, I. W. McKeague, A. Sourander et al. 2014. Increased risk of autism spectrum disorders at short and long interpregnancy intervals. Child Adolesc Psychiatry 53: 1074–81.

[53] Lyall, K., R. J. Schmidt and I. Hertz-Picciotto. 2014. Maternal lifestyle and environmental risk factors for autism spectrum disorders. International Journal of Epidemiology 43(2): 443–464.

[54] Volk, H. E., T. Kerin, F. Lurmann, I. Hertz-Picciotto, R. McConnell and D. B. Campbell. 2014. Autism spectrum disorder: interaction of air pollution with the MET receptor tyrosine kinase gene. Epidemiology 25: 44–7.

[55] Ornoy, A. 2009. Valproic acid in pregnancy: how much are we endangering the embryo and fetus? Reprod Toxicol. 28: 1–10.

[56] Connors, S. L., D. E. Crowell, C. G. Eberhart, J. Copeland, C. J. Newschaffer, S. J. Spence et al. 2005. β_2-adrenergic receptor activation and genetic polymorphisms in autism: data from dizygotic twins. J. Child Neurol. 20: 876–884.

[57] Rossignol, D. A., S. J. Genuis and R. E. Frye. 2014. Environmental toxicants and autism spectrum disorders: a systematic review. Transl Psychiatry 4: e360.

[58] Surén, P., C. Roth, M. Bresnahan, M. Haugen, M. Hornig, D. Hirtz et al. 2013. Association between maternal use of folic acid supplements and risk of autism spectrum disorders in children. J. Am. Med. Assoc. 309: 570–7.

[59] Bleuler, E. 1950 [1911]. Dementia Praecox or the Group of Schizophrenias. New York: International Universities.

[60] Evans, B. 2013. How autism became autism: The radical transformation of a central concept of child development in Britain. History of the Human Sciences 26(3): 3–31.

[61] Kanner, L. 1943. Autistic disturbances of affective contact. Nervous Child 2: 217–250.

[62] Rutter, M. 1968. Concepts of autism: A review of research. Journal of Child Psychology and Psychiatry 9: 1–25.

[63] Asperger, H. 1944/1991. Die autistischen psychopathen im kindesalter. Archiv fur Psychiatrie und Nervenkrankheiten 117: 76–136, 1944, "Autistic psychopathy" in childhood. Translated and annotated by U. Frith. pp. 37–92. *In*: Frith, U. (ed.). Autism and Asperger Syndrome. New York: Cambridge University Press.

[64] Asperger, H. 1979. Problems of infantile autism. Communication 13: 45–52. (Lecture given at Fribourg, Switzerland on March 13, 1977).

Chapter 7
Intellectual Disabilities

Sophia Marra,[1,]* *Emre Tokgoz*[2] and *Percin Karakol*[3]

7.1 Introduction

Intellectual disability is characterized by limitations in cognitive and adaptive function [2]. Current studies suggest that 1 to 3% of the general population have some form of intellectual disability [1]. Our understanding of intellectual disability is constantly being revised and improved. The first mentions of an intellectual disability are traced back to early Egyptian, Greek and Roman civilizations [23]. The first recorded use of the term "mental retardation" was in 1552 B.C. in the Therapeutic Papyrus of Thebes [44]. Unfortunately, even in these ancient civilizations, there was sigma associated disability. Individuals with disabilities were seen as inferior [44]. Throughout history, individuals with intellectual disability have commonly been seen as "other" and a group that should be separate from the rest of society. For example, during the Middle Ages people with disabilities were kept in "idiot cages" in the town square to keep them "out of trouble" [45]. In the 1800s, institutions for individuals with disabilities commonly had poor living conditions and functioned as a means of segregating people with disabilities from the rest of the population [44].

Over the course of hundreds of years, as society has evolved so too has our understanding and researching of intellectual disability [44]. Of note, the terminology used to define intellectual disability has evolved past terms like "mental retardation," "feeble-minded" and "mental handicap." Additionally, there have been legislative initiatives to protect individuals with disabilities. For example, the Americans with Disabilities Act was passed in 1990 to protect the employment and public accommodations of those with disabilities [45]. While there has been progress in amount of research devoted to understanding intellectual disabilities and fighting the

[1] Frank H. Netter School of Medicine, Quinnipiac University, North Haven, CT, 06473, USA.
[2] School of Engineering and Technology, Farmingdale State College, New York, 11735 USA.
[3] Başakşehir Cam and Sakura City Hospital Plastic Reconstructive and Aesthetic Surgery, Istanbul, Turkiye.
* Corresponding author: Sophia.Marra@quinnipiac.edu

DOI: 10.1201/9781003646044-7

stigma against disabilities, there is still a long way to go. This paper summarizes the current state of research on intellectual disability.

7.2 Clinical Presentation

The current prevalence of some forms of intellectual disability is reported to be between 1% to 3% of the population [1]. Intellectual disability (ID) is defined as a condition in which cognitive and adaptive ability are limited [2]. Symptoms occur during the "developmental period" which occurs before 22 years of age [2, 10]. Specifically, the onset of symptoms occurs between infancy through adolescence [1]. Certain disorders have been associated with and commonly co-occur in those with ID. These disorders include depression, anxiety, autism spectrum disorder, attention deficit hyperactivity disorder, infantile cerebral palsy, and meningitis [9].

7.2.1 Adaptive Function

Adaptive functioning refers to the ability to independently carry out tasks and skills required for daily living. These skills commonly fall under the following domains [7, 9]:

1. **Practical skills.** This includes activities of daily living such as grocery shopping, personal care, transportation, tool usage, and household tasks like cleaning, laundry and food preparation.
2. **Socialization skills.** This includes the ability to express one's needs, self-esteem, gullibility, naivety, social problem resolution, adherence to societal rules and interpersonal skills such as proper turn-taking during conversation, and expression recognition.
3. **Conceptual skills.** This includes the ability to understand time, finances and language.

7.2.2 Cognitive Function

Cognitive function is commonly measured using the intelligence quotient (IQ) standardized tests. Cognitive or intelligence functioning includes abilities such as logical reasoning, problem solving, the ability to learn new information, and verbal skills. In general, cognitive function refers to a vast number of skills, capabilities, thoughts, and emotions that allow one to understand and interact within their environment [9].

7.3 Diagnosis

The diagnosis of ID is a multistep process that includes history taking, medical examination, cognitive and adaptive function assessment, comorbid psychiatric disorder identification, and laboratory testing if necessary [14]. This comprehensive approach allows for the proper examination and diagnosis of individuals with intellectual disabilities. During this process, there are certain factors that the medical team must keep in mind.

Specifically, The American Association on Intellectual and Developmental Disabilities (AAIDD) states that ID is defined with five tenants in mind [10]:

1. Functional limitations noted must be made within the context of the abilities of the "individual's age peers and culture".
2. Assessments must take into consideration culture, linguistic, communication, sensory, motor and behavioral differences.
3. Individuals will often have both limitations and strengths.
4. Limitations are identified with the purpose of developing proper support plans.
5. Individualized support plans over a sustained period will generally improve the "life functioning of the person with ID."

7.3.1 Intellectual Functioning Assessment

Diagnosis of ID requires significant deficits in the areas of adaptive and cognitive function. This assessment is done utilizing both clinical assessment and standardized testing. Intellectual function is commonly measured using intelligence quotient scoring. However, the Diagnostic and Statistical Manual of Mental Disorders, 5th edition (DSM-5) strives to decentralized intelligence quotient scoring as the sole diagnosis criteria of ID [15]. This allows an individual's overall abilities to be analyzed without overemphasizing one assessment. DSM-5 defines the IQ cutoff for ID to be 65–75 (IQ of 70 +/– a standard error of 5) [14]. Put differently, ID refers to an IQ score approximately two standard deviations or more below the population [17].

On the other hand, the International Classification of Diseases, 10th revision (ICD-10), urges for a cutoff of 70 IQ score. Additionally, the ICD-10 specifics IQ scores for four severity levels of ID: mild (IQ 50 to 69), moderate (IQ 35 to 49), severe (20 to 34) and profound (IQ less than 20) [14]. The IQ test is not designed to be given to children under the age of 3 years old [14]. In this population and those unable to complete the exam, comparable standardized assessments may be administered.

7.3.2 Adaptive Functioning Assessment

Adaptive function is examined using standardized assessment tools. Specifically, they aim to assess the three domains of adaptive function: practical, social and conceptual skills. Commonly information is collected from someone close to the individual such as a parent, guardian or teacher by a licensed psychologist [16]. There are multiple assessment tools utilized by medical professionals. As mentioned, diagnosis of ID is multifactorial and should be individualized based on the patient's needs and abilities. Current assessment tools include [16]:

1. **Adaptive Behavior Evaluation Scale (ABES).** This test is used to assess individuals with already-defined mental or developmental disabilities. Assessment subscales include areas of daily living, communication skills and behavior regulation.

2. **Adaptive Behavior Assessment System (ABAS).** This assessment can be used in both pediatric and adult populations. Questions assess appropriate daily living skills based on one's age.
3. **Vineland Adaptive Behavior Scales (VABS).** This tool is validated for individuals from the time of birth to the age of 19 years old. There are four subsections that assess communication, daily living, socialization, problem-solving and motor functioning skills.
4. **The Diagnostic Adaptive Behavior Scale (DABS).** This assessment was developed by the AAIDD to analyze all three areas of adaptive behavior including practical, social, and conceptual skills. DABS specifically aims to determine the best support for individuals to maintain the highest level of independent living. This tool is validated in individuals 4 to 21 years old.

7.3.3 Age of Onset Criteria

The age of onset is also a critical element of ID diagnosis. This criterion is based on findings that suggest that ID develops during the "developmental period." This is a timeframe in which brain development is rampant and critical. Typically, the developmental period is thought to end around 22 years of age [10]. Current literature shows that there is crucial brain growth and maturation up to approximately the age of 22 years old, specifically concerning cortical gray matter volume, corpus callosum, and white matter [11, 12, 13].

7.3.4 Levels of Severity

As mentioned, IQ must be evaluated in the wider context of the individual's strengths and limitations. Below are additional characteristics attributed to the four severities of ID [1]:

1. **Mild.** Mild ID comprises 85% of people with intellectual disability and diagnosis does not occur before the age of 5–6 years old [19, 20, 21]. Individuals with mild ID may show limitations with language, arithmetic, writing and other academic skills [3, 20, 21, 22]. Additionally, those with mild ID may struggle with short-term recall, executive functioning and understanding social context and interactions [3]. Persons with mild ID may exhibit academic abilities in line with a 4th to 5th grade education level [21, 22]. Individuals with mild ID are often able to successfully complete most daily activities of life and lead independent lives as adults [3, 20, 22]. Some assistance may be valuable in areas of healthcare, financial management and legal decisions [3, 20, 21, 22].
2. **Moderate.** Moderate severity ID comprises about 10% of all cases of ID and diagnosis usually occurs by the age of 3–5 years old [19, 20, 21]. Common initial signs of moderate ID are the delayed acquisition of language as a child [19, 21, 22]. Individuals with moderate ID may struggle with reading, writing, mathematics, and understanding social context and norms [3, 20, 22]. Individuals with moderate ID commonly exhibit the academic skills of an early elementary school grade education [3, 22]. Adults with moderate ID are often

able to maintain independence in daily activities of life with the help of effective support systems [20].

3. **Severe.** Severe ID comprises about 4% of all cases of ID and diagnosis usually occurs before the age of 3 years old [19, 20, 21]. Individuals with severe ID often exhibit limitations in spoken language, social interactions, concepts of time, quantity and time, problem solving, and financial management [3, 20, 21, 22]. Assistive devices can be implemented to facilitate skills such as communication [1, 3, 22]. Communication may also include the use of single words or phrases [3, 22]. Those with severe ID may exhibit the cognitive skills of a child around 3 to 5 years old [3, 19, 21, 22]. In addition to cognitive limitations, those with severe ID may also have motor deficits as well as other conditions further contributing to intellectual and adaptive function limitations [3, 19]. Individuals with severe ID commonly benefit from significant, consistent support in many or all facets of daily activities of life [3, 22].
4. **Profound.** Profound ID comprises about 1% of all cases of ID and diagnosis usually occurs before the age of 2 years old [19, 20, 21]. Profound ID is differentiated from severe ID using assessments of adaptive function because intelligence assessments are not validated in persons with intellectual functioning below the 0.003rd percentile [22]. Individuals with profound ID may experience significant limitations in academics, conceptualization, communication, continence, movement, and self-care [1]. Those with profound ID may exhibit the cognitive skills of child around the age of 3 years old [3, 19, 21, 22]. Persons with profound ID commonly benefit from total care support for activities of daily living [3, 20].

7.4 Etiology

Intellectual disabilities are a multifactorial and complex disability with many different possible etiologies. Persons with severe ID are more likely than those with mild ID to have the cause of their intellectual disability identified [1]. The etiology of intellectual disability can broadly be separated into genetic and environmental causes [1].

7.4.1 Genetics

Intellectual disabilities can trace their origins to both genetic and non-genetic origins. Current literature suggests that up to 40% of all intellectual and developmental disabilities have genetic causes [23]. As mentioned above, it is more common to identify the etiology in severe cases of ID compared to patients with more mild presentations of ID. The results of genetic testing can afford patients and their families a better understanding of their ID diagnosis and inform future clinical steps as well as potential comorbidities [18].

7.4.1.1 Single Gene

One potential cause of ID is single gene mutations. Single gene disorders can lead to ineffective enzymatic activity which then interferes with normal metabolic pathways [26, 27]. As a result, there is an accumulation of toxic byproducts. These toxic byproducts can then cause intellectual disabilities [26, 27]. Single gene mutations can be inherited as X-linked, autosomal recessive or autosomal dominant traits. Out of all single gene metabolic disorders, autosomal recessive is most common [18]. One example of autosomal recessive disorder that can result in intellectual disability is phenylketonuria (PKU). PKU results in the accumulation of phenylalanine which can cause cognitive impairments in addition to epilepsy [9].

About 1 to 3% of the general population is affected by X-linked intellectual disabilities (XLID) [28]. Males are more likely than females to inherited X-linked mutations [18]. Currently, there are over 160 X-linked genes that have been shown to be associated with intellectual disabilities [24]. Many of these genetic mutations can result in not only intellectual disability but also disorders such as epilepsy and physical malformations [9, 18, 29, 30]. For example, one of the most common XLIDs is Fragile X syndrome [32]. Symptoms usually include intellectual disability as well as distinct physical characteristics such as a narrow face, large ears and head, and flexible joints [33]. Another example is Alpha thalassemia X-linked intellectual disability (ATR-X) syndrome. Here the ATRX gene, found on the X-chromosome, is mutated resulting in ATR-X syndrome [29]. In addition to symptoms of ID, people with ATR-X may also have hypotonia and physical malformations of the head, genitals and/or skull [25, 29].

7.4.1.2 Chromosomal Abnormality

Chromosomal abnormalities leading to ID can take various forms such as copy number variants (CNV), aneuploidy and epigenetics. Aneuploidy is defined as the existence of an extra or missing chromosome. The most common example of aneuploidy that results in intellectual disability is Trisomy 21, otherwise known as Down Syndrome. CNV mutations, on the other hand, refer to the microdeletion or duplication of genetic material, but not the entire chromosome. One example of a CNV mutation that results in ID is repetitive deletions at position 15q11-q13. CNVs in this genomic area have been associated with both Prader Willi and Angelman's syndrome. Both syndromes are associated with intellectual disability [31].

7.4.2 Environmental

In addition to genetic causes, there are also environmental causes of intellectual disabilities. One of the most common non-genetic causes of ID is infection and intoxicants. Specifically, meningitis has been associated with ID development. Meningitis is a viral infection of the meninges that may result from childhood illnesses like the chickenpox or mumps. Lead poisoning can also cause intellectual disability as well as epilepsy and central nervous system damage. This is commonly a risk for young children and infants living in older homes that may potentially ingest lead-based paint chips. Another non-genetic cause of ID is *in utero* toxicity [32].

In utero toxicity refers to the introduction of teratogens to the fetal environment during pregnancy. The most common example of fetal toxicity leading to ID is fetal alcohol syndrome (FAS) [32]. Fetal alcohol syndrome falls under the umbrella of fetal alcohol spectrum disorders [35]. FAS is the result of maternal consumption of alcohol which interferes with fetal neurological development [35]. FAS symptoms include intellectual disability, small stature, and facial abnormalities including small eyes, thin upper lip and a smooth transition from nose to upper lip [34].

7.5 Comorbidities

Many individuals with intellectual disabilities also live with medical comorbidities. Additionally, many conditions have differing prevalences in people with moderate to severe ID compared to mild ID.

7.5.1 Comorbidity with Autism

Autism and Intellectual Disabilities are recognized as two distinct disabilities, yet they are often co-occurring [3]. Current data from the Centers for Disease Control and Prevention's (CDC) Autism and Developmental Disabilities Monitoring (ADDM) Network, suggest that the prevalence of children with co-occurring ID and autism is 48.5% at 4 years old, 37.9% at 8 years old, and 36.6% at 16 years old [4, 5, 6]. Studies also show that the frequency of co-occurring ID and autism is significant higher in Non-Hispanic Black children and lower in Hispanic children compared to Non-Hispanic White children [8]. Current research suggests that 30% of intellectual disabilities and autism are genetic in origin [42]. Up to 2000 genes, many found on the X chromosome, have been shown to be associated with co-occurring intellectual disability and autism spectrum disorder [42].

7.5.2 Comorbidity with Epilepsy

Epilepsy has been shown to be associated with intellectual disabilities that are genetic in origin. More than 5% of the general population have both epilepsy and intellectual disability. Additionally, 20% of individuals with intellectual disability also have epilepsy. Current literature suggests that the association between these two disorders is rooted in mutations of proteins responsible for DNA translation as well as mRNA processing, modulation and stability. The brain relies on such proteins to maintain proper neuronal excitability- a process that malfunctions in epilepsy. These proteins also allow for creation of new synaptic connections to allow for learning and memory- two of the major deficits associated with intellectual disability. Activity-dependent translation is critical for proper neurological function and thus deficits in this process preferential lead to epilepsy with intellectual disabilities [37].

7.5.3 Comorbidity with Psychiatric Disorders

Current literature suggests that from 10% to 60% of individuals with ID also have a mental disorder [19]. The great deal of variation is likely due to differing assessment tools and methodology. Additionally, individuals with intellectual disabilities have double the prevalence of mental health problems compared to those without intellectual disabilities [40]. Studies have even shown that 29% of individuals with intellectual disabilities have an undiagnosed mental disorder [38]. The two most common disorders are major depressive and anxiety disorders [38]. Individuals with mild intellectual disabilities also have a higher prevalence of schizophrenia compared to those without intellectual disability [41].

7.6 Future Directions

While there have been advancements in our understanding of intellectual disability, we propose the following suggestions for future directions. First, there is very limited representation in the research of intellectual disability. Few studies report data that has focuses on underrepresented groups such as females, non-White individuals, and those of lower economic status. For example, current research focused on autism has shown that females likely have a different constellation of symptoms than males which may greatly impact the number of females that are diagnosed with autism. Additionally, a lack of understanding about how autism may present differently in various groups limits the quality of individualized care that can be given. With that being said, the association between autism and intellectual disability is well documented, so it is important that we understand how ID may be present differently in various groups for the same reason.

Secondly, current data shows that those with milder forms of ID are less likely to know the etiology of their disorder. Furthermore, there is rather limited information in general on the etiology of ID. Intellectual disabilities are a heterogenous disability that can be the result of many different causes. With current advancements in areas of genetic testing, it will be very important to focus on understanding what the biological underpinnings of the various forms of ID are to best treat each patient. Individualized medicine is becoming more commonplace and accessible. Every person with intellectual disabilities is unique and has their own individual strength, limitations, and needs. More accurate etiological information will help patients better understand their diagnosis and propel more effective care.

Finally, comorbidities are very common within intellectual disabilities. As previously mentioned, ID is very heterogenous. With that, there is a great deal of interplay with other disorders. Some of the most researched comorbidities are outlined in this review. But there is still a great deal of unknown when it comes to diagnosing comorbid disorders-especially mental disorders. In addition, future research would benefit from focusing on how and why comorbid disorders are associated with ID and what common pathways they share. This is again helping patients receive the most effective, personalized care.

References

[1] Patel, D. R., M. D. Cabral, A. Ho and J. Merrick. 2020. A clinical primer on intellectual disability. Translational Pediatrics 9: S23.

[2] Patel, D. R., D. E. Greydanus, J. Merrick and I. L. Rubin. 2016. Introduction to intellectual and developmental disabilities. Health Care for People with Intellectual and Developmental Disabilities Across the Lifespan 5–12.

[3] American Psychiatric Association. 2013. Diagnostic and Statistical Manual of Mental Disorders (5th ed.). American Psychiatric Association. https://doi.org/10.1176/appi.books.9780890425596.
[4] Hughes, M. M., K. A. Shaw, M. E. Patrick, M. DiRienzo, A. V. Bakian, D. A. Bilder et al. 2023. Adolescents with autism spectrum disorder: diagnostic patterns, co-occurring conditions, and transition planning. Journal of Adolescent Health 73(2): 271–278.
[5] Maenner, M. J. 2023. Prevalence and characteristics of autism spectrum disorder among children aged 8 years—Autism and Developmental Disabilities Monitoring Network, 11 sites, United States, 2020. MMWR. Surveillance Summaries, 72.
[6] Shaw, K. A. 2023. Early identification of autism spectrum disorder among children aged 4 years—Autism and Developmental Disabilities Monitoring Network, 11 sites, United States, 2020. MMWR. Surveillance Summaries, 72.
[7] McQuaid, G. A., K. A. Pelphrey, S. Y. Bookheimer, M. Dapretto, S. J. Webb, R. A. Bernier et al. 2021. The gap between IQ and adaptive functioning in autism spectrum disorder: Disentangling diagnostic and sex differences. Autism 25(6): 1565–1579.
[8] Furnier, S. M., R. Gangnon, J. L. Daniels, S. Ellis Weismer, C. Nadler, K. Pazol et al. 2024. Racial and ethnic disparities in the co-occurrence of intellectual disability and autism: Impact of incorporating measures of adaptive functioning. Autism Research 17(3): 650–667.
[9] Lee, K., M. Cascella and R. Marwaha. 2019. Intellectual disability.
[10] Schalock, R. L., R. Luckasson and M. J. Tassé. 2021. An overview of intellectual disability: Definition, diagnosis, classification, and systems of supports. American Journal on Intellectual and Developmental Disabilities 126(6): 439–442.
[11] Jiang, X. and J. Nardelli. 2016. Cellular and molecular introduction to brain development. Neurobiology of Disease 92: 3–17.
[12] Mills, K. L., F. Lalonde, L. S. Clasen, J. N. Giedd and S. J. Blakemore. 2014. Developmental changes in the structure of the social brain in late childhood and adolescence. Social Cognitive and Affective Neuroscience 9(1): 123–131.
[13] Giedd, J. N., J. Blumenthal, N. O. Jeffries, F. X. Castellanos, H. Liu, A. Zijdenbos et al. 1999. Brain development during childhood and adolescence: a longitudinal MRI study. Nature Neuroscience 2(10): 861–863.
[14] Kishore, M. T., G. A. Udipi and S. P. Seshadri. 2019. Clinical practice guidelines for assessment and management of intellectual disability. Indian Journal of Psychiatry 61(Suppl 2): 194–210.
[15] American Psychiatry Association. Intellectual disability. Accessed December 28, 2025. https://www.psychiatry.org/File%20Library/Psychiatrists/Practice/DSM/APA_DSM-5-Intellectual-Disability.pdf.
[16] (ECCM), E. C. C. M. The basics of adaptive functioning assessments - ECCM. https://www.eccm.org/blog/the-basics-of-adaptive-funtioning-assessments.
[17] Caemmerer, J. M., S. R. Young, E. Blemahdoo and E. L. Winter. 2023. Intelligence tests. pp. 327–346. In Handbook of Clinical Child Psychology: Integrating Theory and Research into Practice. Cham: Springer International Publishing.
[18] Somanadhan, S., N. L. Johnson, B. S. Gilroy, A. Lawlor and J. Vockley. 2023. Intellectual and developmental disabilities and rare diseases. pp. 89–104. In Intellectual Disabilities: Health and Social Care Across the Lifespan. Cham: Springer International Publishing.
[19] Harris, J. C. 2006. Intellectual Disability: Understanding Its Development, Causes, Classification, Evaluation, and Treatment. Oxford University Press.
[20] Schalock, R. L. 2010. Intellectual disability. Cross-Cultural Psychology: Contemporary Themes and Perspectives 312.
[21] AlSalehi, S. M. and E. H. Alhifthy. 2020. Developmental delay and intellectual disability. Clinical Child Neurology 237–256.
[22] World Health Organization. Disorders of intellectual development. https://icd.who.int/browse11/l-m/en#/http%3a%2f%2fid.who.int%2ficd%2fentity%2f605267007.
[23] Kaufman, L., M. Ayub and J. B. Vincent. 2010. The genetic basis of non-syndromic intellectual disability: a review. Journal of Neurodevelopmental Disorders 2: 182–209.
[24] Schwartz, C. E., R. J. Louie, A. Toutain, C. Skinner, M. J. Friez and R. E. Stevenson. 2023. X-Linked intellectual disability update 2022. American Journal of Medical Genetics Part A 191(1): 144–159.

[25] Chen, H. 2017. Alpha-Thalassemia X-Linked Mental Retardation Syndrome. *In*: Atlas of Genetic Diagnosis and Counseling. Springer, New York, NY. https://doi.org/10.1007/978-1-4939-2401-1_9.

[26] Guerrero, R. B., D. Salazar and P. Tanpaiboon. 2018. Laboratory diagnostic approaches in metabolic disorders. Annals of Translational Medicine 6(24).

[27] Gilbert-Barness, E. and P. M. Farrell. 2016. Approach to diagnosis of metabolic diseases. Translational Science of Rare Diseases 1(1): 3–22.

[28] Bassani, S., J. Zapata, L. Gerosa, E. Moretto, L. Murru and M. Passafaro. 2013. The neurobiology of X-linked intellectual disability. The Neuroscientist 19(5): 541–552.

[29] Chang, C. W., M. Li, X. Xu and Y. Liu. 2019. Chapter 11-Helicase dysfunctions in human diseases. Helicases from All Domains of Life, R. Tuteja, Editor, 191–210.

[30] Mullen, S. A., G. L. Carvill, S. Bellows, M. A. Bayly, S. F. Berkovic, L. M. Dibbens et al. 2013. Copy number variants are frequent in genetic generalized epilepsy with intellectual disability. Neurology 81(17): 1507–1514.

[31] Wolfe, K., A. Strydom, D. Morrogh, J. Carter, P. Cutajar, M. Eyeoyibo et al. 2017. Chromosomal microarray testing in adults with intellectual disability presenting with comorbid psychiatric disorders. European Journal of Human Genetics 25(1): 66–72.

[32] Shree, A. and P. C. Shukla. 2016. Intellectual Disability: Definition, classification, causes and characteristics. Learning Community-An International Journal of Educational and Social Development 7(1): 9–20.

[33] U.S. Department of Health and Human Services. (n.d.-b). What are the symptoms of fragile X syndrome?. Eunice Kennedy Shriver National Institute of Child Health and Human Development. https://www.nichd.nih.gov/health/topics/fragilex/conditioninfo/commonsymptoms#:~:text=When%20these%20children%20start%20to,feet%2C%20and%20a%20prominent%20forehead.

[34] Professional, C. C. medical. (n.d.). Fetal alcohol syndrome (FAS): Symptoms, causes and treatment. Cleveland Clinic. https://my.clevelandclinic.org/health/diseases/15677-fetal-alcohol-syndrome.

[35] Vorgias, D., F. D. Bynum and B. Bernstein. 2023. Fetal alcohol syndrome. In StatPearls [Internet]. StatPearls Publishing.

[36] Uzun Cıcek, A., S. A. Sarı and C. Mercan Isık. 2020. Sociodemographic characteristics, risk factors, and prevalence of comorbidity among children and adolescents with intellectual disability: a cross-sectional study. Journal of Mental Health Research in Intellectual Disabilities 13(2): 66–85.

[37] Malone, T. J. and L. K. Kaczmarek. 2022. The role of altered translation in intellectual disability and epilepsy. Progress in Neurobiology 213: 102267.

[38] Pena-Salazar, C., F. Arrufat, J. M. Santos, A. Fontanet, G. González-Castro, S. Más et al. 2020. Underdiagnosis of psychiatric disorders in people with intellectual disabilities: Differences between psychiatric disorders and challenging behaviour. Journal of Intellectual Disabilities 24(3): 326–338.

[39] Koskentausta, T., M. Iivanainen and F. Almqvist. 2007. Risk factors for psychiatric disturbance in children with intellectual disability. Journal of Intellectual Disability Research: JIDR 51(1): 43–53.

[40] Totsika, V., A. Liew, M. Absoud, C. Adnams and E. Emerson. 2022. Mental health problems in children with intellectual disability. The Lancet Child and Adolescent Health 6(6): 432–444.

[41] Welch, K. A., S. M. Lawrie, W. Muir and E. C. Johnstone. 2011. Systematic review of the clinical presentation of schizophrenia in intellectual disability. Journal of Psychopathology and Behavioral Assessment 33: 246–253.

[42] Chiurazzi, P., A. K. Kiani, J. Miertus, S. Barati, E. Manara, S. Paolacci et al. 2020. Genetic analysis of intellectual disability and autism. Acta Bio Medica: Atenei Parmensis 91(Suppl 13).

[43] Roth, E. A., S. N. Sarawgi and J. C. Fodstad. 2019. History of intellectual disabilities. Handbook of Intellectual Disabilities: Integrating Theory, Research, and Practice 3–16.

[44] The Minnesota Governor's Council on Devlopmental Disabilities. (n.d.-b). Parallels in timea history of developmental disabilities. Parallels In Time | A History of Developmental Disabilities | Part One. https://mn.gov/mnddc/parallels/two/2.html.

[45] Disabilities and the law: The evolution of independence. (n.d.-a). https://www.fedbar.org/wp-content/uploads/2010/09/disabilitiesandthelaw-September2010-pdf-1.pdf.

Chapter 8

Attention Deficit/ Hyperactivity Disorder

Sophia Marra,[1,*] *Percin Karakol*[2] and *Emre Tokgoz*[3]

8.1 Introduction

Schizophrenia is characterized by hallmark symptoms of psychosis and decline in function [1]. Schizophrenia, along with additional psychiatric disorders, has been documented throughout history but without clear understandings of the disorder [17]. Current strides have been made to identify the prevalence, risk factors, presentation, pathology and etiology of schizophrenia. The development of the Diagnostic and Statistical Manual of Mental Disorders (DSM) has helped to standardize research, clinical care, and the general conversation around schizophrenia and many mental disorders [17]. Before the inception of DSM, psychosis – one of the hallmark symptoms of schizophrenia – was separated into two categories: manic depressive psychosis and dementia praecox [96]. Emil Kraepelin described dementia praecox, or dementia of the young, as having both positive and negative symptoms [96]. Eugen Bleuler was the first to use the term "schizophrenia" and diagnose individuals with schizophrenia [97]. He detailed the disorder as involving thought, emotion and behavior. Additionally, Bleuler recognized the chronic timeline of the disorder with periods of acute exacerbations of symptoms [97]. Since then, the definition, diagnostic criteria, etiology, clinical presentation, pathology, and risk factors of schizophrenia have been further developed and researched. This article highlights the most current advancements in the field.

[1] Frank H. Netter School of Medicine, Quinnipiac University, North Haven, CT, 06473, USA.
[2] Başakşehir Cam and Sakura City Hospital Plastic Reconstructive and Aesthetic Surgery, Istanbul, Turkiye.
[3] School of Engineering and Technology, Farmingdale State College, New York, 11735 USA.
* Corresponding author: Sophia.Marra@quinnipiac.edu

DOI: 10.1201/9781003646044-8

8.2 Clinical Presentation

Schizophrenia is a disorder characterized by psychosis and functional decline [1]. Studies suggest that around 20.9 million people across the world have schizophrenia [2]. Although the prevalence is relatively low, schizophrenia is one of the main contributors to global burden of disease [4, 5, 6]. This can be the result of multiple factors including the severity and persistence of the disorder, even when given optimal treatment options [6]. Additionally, there are great financial costs that can be associated with schizophrenia [5]. The life expectancy of individuals with schizophrenia is 15 years less than those without schizophrenia [3]. Even though this disorder has been researched for many years, there is still a great deal that we do not understand. There is a global demand to better understand schizophrenia to better serve patients and lessen the burden of the disease.

8.2.1 Symptoms

Schizophrenia is diagnosed at the onset of psychosis symptoms. Some studies suggest however that those that are eventually diagnosed with schizophrenia exhibited certain characteristics in childhood, such as social difficulties or poor motor control [24, 25]. Traditionally, the symptoms of schizophrenia have been organized into three categories: positive, negative and general psychopathology [19]. This is based on the model put forth by Kay et al. in 1987 called the Positive and Negative Syndrome Scale (PANSS) [20].

Positive symptoms describe those that are a heightened or more intense version of normal function. Some examples include [20, 23, 27]:

- **Hallucination.** The perception of auditory, tactile or visual sensation without external stimuli. Auditory hallucinations are the most common in individuals with schizophrenia.
- **Delusion.** Beliefs held even in the face of evidence to the contrary
- **Disorganized thinking.** Characterized by "incoherence, derailment or illogicality" [21]
- **Excitement.** Can manifest as hyperactivity, fast motor activity, or hypervigilance
- **Grandiosity.** Exaggerated self-image or feelings of self-importance, superiority
- **Suspiciousness.** Exaggerated beliefs of distrust or persecution
- **Hostility.** Verbal and non-verbal resentment or anger [26]

Whereas negative symptoms are characterized by a general lack of functioning in areas of cognition, emotions, and social interaction [20]. Some examples include [20, 23, 27]:

- **Blunted affect.** Lack of emotional responsiveness
- **Emotional withdrawal.** Lack of interest or engagement in activities of life
- **Poor rapport.** Lack of expression of empathy or openness with others

- **Passive-apathetic social withdrawal.** Lack of interest in social relationships due to passivity or apathy
- **Difficulty in abstract thinking.** Lack of ability to effectively engage in problem solving tasks
- **Lack of spontaneity and flow of conversation.** Lack of normal flow in conversation
- **Stereotyped thinking.** Increased rigidity and repetitive thought

General psychopathology symptoms include [20, 23, 27]:

- **Somatic concern.** Beliefs of physical/bodily illness or injury
- **Anxiety.** Increased nervousness and worry
- **Guilt feelings.** Can be remorse for real or imagined misdeeds
- **Tension.** Physical manifestation of anxiety and agitation including stiffness, tremor, sweating or restlessness
- **Mannerisms and posturing.** Unnatural movements that tend to be awkward or disorganized
- **Depression.** Feelings are sadness or helplessness
- **Motor retardation.** Diminished or slowed movement and speech, reduced muscle tone, and responsiveness to stimuli
- **Uncooperativeness.** Refusal to comply with others including interviewer, family, hospital staff
- **Unusual thought content.** Thoughts that are characterized as "strange", fantastical, or "bizarre"
- **Disorientation.** Reduced awareness of one's environment or surroundings
- **Poor attention.** Reduced ability to focus alertness
- **Lack of judgment and insight.** Reduced understanding of one's own psychiatric condition
- **Disturbance of volution.** Lack of faculty to control one's behaviors, actions and words
- **Poor impulse control.** Reduced ability to control sudden urges
- **Preoccupation.** Increased focus placed on internal thoughts and feelings
- **Active social avoidance.** Reduced involvement in social interaction

Schizophrenia symptoms are also categorized as: hallucinations/psychosis, negative symptoms and cognitive symptoms [22]. Cognitive symptoms refer to those that impact memory and executive function [25]. The same characteristic symptoms of schizophrenia are included in either organization of said symptoms. Interestingly, the Diagnostic and Statistical Manual of Mental Disorders, 5th edition significantly emphasizes positive symptoms. Negative symptoms are mentioned once, and cognitive symptoms are not mentioned at all [25].

8.2.2 Onset of Schizophrenia

Some research suggests that schizophrenia commonly has its onset in adolescence [7]. However, there is a great deal of literature that has focused on identifying and defining the onset of schizophrenia. Overall, there is a general consensus on the existence of the following categories of schizophrenia onset [8]:

1. *Adult-Onset Schizophrenia (AOS)*- Schizophrenia symptom onset is at the age of 18 years old or older
2. *Early Onset Schizophrenia (EOS)*- Schizophrenia symptom onset is before the age of 18 years old
   a. *Very Early Onset Schizophrenia (VEOS)*- Schizophrenia symptoms onset is specifically before the age of 13 years old

There are important characteristics associated with the timeframe that symptoms begin [16]. Individuals with an earlier onset of schizophrenia tend to have:

- More severe symptoms [15]
- More negative symptoms [10]
- Poorer prognosis [15]
- More misdiagnoses [16]
- More deficits before schizophrenia onset [11, 12]
- Higher levels of urinary incontinence [12]
- Higher levels of characteristics of autism [13]
- Longer duration of untreated psychosis (DUP) [15]
- Higher suicidality and more depressive symptoms [14]

Additionally, studies suggest that 4% of all schizophrenia cases begin before the age of 15 years old [9]. Furthermore, 0.5–1% begin before the age of 10 years old [9]. The findings above highlight the importance of effective diagnostic criteria, support and treatment options for all individuals with schizophrenia.

8.2.3 Timeline of Schizophrenia

There are many models that have been put forth regarding the timeline of schizophrenia. Current literature suggests that schizophrenia symptoms most often start with an initial psychotic episode. This is considered the beginning of Stage 1 [41]. Before this point, the individual is said to be in the prodrome phase or Stage 0.

8.2.3.1 Prodromal Phase- Stage 0

The prodromal phase refers to the period before the initial psychotic episode. This stage can be characterized by mild to moderate symptoms including depression, lack of motivation and changes in vision, attitude or mindset [58, 59]. The timeframe of the prodromal phase can range from weeks to years but most often lasts one year

[60]. This phase is only truly identified retroactively, after the individual has entered the succeeding stages and been diagnosed with schizophrenia [60].

8.2.3.2 Acute Manifestation of Psychotic Symptoms- Stage 1

The acute phase is characterized by an initial episode of psychotic symptoms commonly triggered by a stressor, whether psychosocial or otherwise [41, 57]. Individuals are more likely to experience positive symptoms including delusions, paranoia and hallucinations [58, 61]. Additionally, individuals may not realize that they are experiencing such symptoms [58, 61, 62]. This denial could lead to symptoms of aggression or irritability [62]. The timeframe of this stage is debated, but some suggest that this acute phase ends within 5 years of first psychotic symptoms [63]. Researchers suggest that after this initial psychotic episode, one third of individuals will enter remission and never have another episode again, and another third of individuals will have one or more future psychotic events; This is characterized as stage 2. The final third of individuals will have several recurrent episodes and have been classified as chronic cases [64].

8.2.3.3 Remission and Relapse- Stage 2

This can be a highly variable time period within each individual and between individuals [41]. Remission describes a phase when symptoms are moderate, and the individual is no longer is an "active" episode of psychosis [65]. Symptom severity is often determined using the PANSS scale outlined above [66]. Relapse is then characterized by a resurgence of symptoms and increased severity that usually impacts one's ability to function in their daily life [67]. Depressed and negative symptoms are most often seen during stage 2 and can greatly impact an individual's quality of life [68]. During relapse, positive symptoms may become more prevalent [67, 69]. The duration of this stage is variable, but some report that this phase lasts an average of 9 years [69].

8.2.3.4 Chronic Psychosis- Stage 3

Chronic psychosis describes the phase where symptoms are persistent and severe [70]. In this stage, individuals may experience primarily depressed, negative and cognitive symptoms [69, 70]. Individuals in stage 3 are at a higher risk of experiencing suicidal thoughts [71]. This phase usually starts 15–20 years after the onset of acute stage psychosis [41].

8.3 Diagnostic Criteria

The publication of the Diagnostic and Statistical Manual of Mental Disorders, 5th edition ushered an updated conceptualization of schizophrenia. The DSM-5 introduced the idea of schizophrenia as being on a spectrum rather than categorical

[17]. Currently, the diagnosis of schizophrenia is made using the following criteria [18]:

- **Criterion A: Presence of symptoms.** Two of more of the following symptoms must be present for the significant portion of a one-month period:
  1. Delusions
  2. Hallucinations
  3. Disorganized Speech
  4. Grossly disorganized or catatonic behavior
  5. Negative symptoms

At least one of the symptoms must include 1–3 (delusions, hallucinations, and/or disorganized speech).

- **Criterion B: Duration of time.** There must be continuous signs of disturbance for at least 6 months. During this period, there must be at least one month of symptoms that meet Criterion A. Additionally, there may be period of prodromal and/or residual symptoms. "Disturbances" outside of the one month of Criterion A symptoms, may take the form of only negative symptoms. Or one might exhibit two or more Criterion A symptoms but at a lesser severity.
- **Criterion C: Social and Occupational Dysfunction.** There is a significant decline in functioning compared to before the onset of the disorder. Areas of functioning can include work, academics, interpersonal relationships, or self-care. In children or adolescents, functioning should be compared to what is expected for their age group.
- **Criterion D: Exclusion of Schizoaffective and Major Mood Disorder.** Schizoaffective and depressive or bipolar disorder must be excluded as possible diagnoses for the observed symptoms. These disorders can be excluded if:
  1. No major depressive or manic episode symptoms are present during active phase (Criterion A) symptoms
  2. Mood episode symptoms occur during active phase symptoms, but the duration is short compared to the duration of active and residual periods.
- **Criterion E: Exclusion of Substance or General Mood Conditions.** Observed symptoms are not the resulting effects of a substance or another condition.
- **Criterion F: Relationship to Global Developmental Delay or Autism Spectrum Disorder (ASD).** Schizophrenia is only diagnosed in individuals with ASD or other childhood-onset communication disorders if there is a presence of delusions or hallucinations for at least 1 month.

8.4 Pathology/Etiology

8.4.1 Neuropathology Overview

Current literature has struggled to identify clear neuropathological characteristics of schizophrenia to the same granularity of a disorder like Alzheimer's Disease with its well-defined amyloid plaques and tau tangles. However, that doesn't mean that we know nothing about the neurological underpinnings of schizophrenia. For example, unlike Alzheimer's Disease there is a consensus that schizophrenia is not characterized by neurodegenerative processes [29].

8.4.1.1 Postmortem Studies

The study of postmortem brain tissue has illuminated many characteristics of schizophrenia. One of the most well-documented and consistent findings is that the parvalbumin (PV) GABAergic interneurons are damaged in schizophrenia; These neurons are responsible for providing recurrent inhibition to pyramidal neurons [30, 31]. Additionally, PV, glutamate decarboxylase (GAD67) and GABA transporter are all downregulated in schizophrenia while the postsynaptic GABAa receptor is upregulated [25, 30, 31]. As a result of this decrease in PV and GABAergic neuronal function, there is a decrease in theta and gamma oscillations which has been associated with the cognitive deficits seen in schizophrenia [32].

Another postmortem study finding is the decrease in dendritic complexity and dendritic spine density in those with schizophrenia compared to controls [25]. The areas of the brain most affected seem to be the dorsal-lateral prefrontal cortex, sensory cortex, and hippocampus [33, 34, 35]. Furthermore, these findings suggest that there is a 25–40% decrease in total cortical pyramidal neuron glutamatergic synapses [25].

8.4.1.2 Brain Imaging

Another critical assessment tool that has increased our understanding of schizophrenia is brain imaging technology. Of note, computer assisted tomography (CAT) scans have shown significant ventricular enlargement in individuals with schizophrenia [36]. Many studies have corroborated this finding, with some reporting ventricular enlargement using ventricular brain ratio (VBR) [37]. While this difference may be statistically significant, it appears to be clinically small, thus making the VBR measurement not diagnostically useful [25].

While the ventricles of individuals with schizophrenia seem to expand, the volume of the temporal and prefrontal cortex seem to decrease [25]. Other studies show that this cortical reduction may involve more areas that just the temporal and prefrontal cortex [38]. But as mentioned above, schizophrenia is not thought to be a neurodegenerative disease [29]. This is because those will schizophrenia lack the biomarkers characteristic of neurodegeneration [25]. Thus, there are a few proposed explanations for this cortical volume reduction observed. One such explanation is that increased cell packing and decreased neuronal somal size, dendritic complexity and spines density that is causing reduced cortical volume in the cortex and hippocampus [33, 35, 39]. Additionally, some research suggests that it is antipsychotic treatment itself rather than schizophrenia that is causing a loss of gray matter [40].

8.4.2 Etiology

8.4.2.1 Dopamine Hypothesis

One of the most common hypotheses for the origin of schizophrenia is the "dopamine hypothesis." This hypothesis was first discovered is an effort to understand exactly antipsychotic medications like chlorpromazine exerted their effects [25]. Arvid Carlsson suggested that these drugs worked by blocking forebrain dopaminergic receptors [42]. This was because these therapeutic agents in some patients produced Parkinsonian-like effects, which is characterized by a reduction in dopaminergic neurons [25, 42]. Specifically, the literature showed that it was D2 dopaminergic neurons that were blocked by these drugs [43]. Thus, the hypothesis of excess dopamine as an underlying mechanism of schizophrenia was born. An excess of dopamine released onto D2 receptors can result in psychosis [22]. As previously mentioned, psychosis is commonly the first symptom at the onset of schizophrenia. Additionally, antipsychotic drugs can block D2 receptors to try to mitigate the effects of excess dopamine [22]. However, the negative and cognitive symptoms of schizophrenia are not affected by antipsychotic drugs [25]. This points to the fact that while the "dopamine hypothesis" is illuminated, there is likely more to the story of schizophrenia.

Since the proposition of the dopamine hypothesis, researchers have investigated the root of the negative and cognitive symptoms of schizophrenia. As a result, the updated version of this hypothesis does propose an explanation for these symptoms. While the positive symptoms of psychosis are related to excess dopamine acting on D2 receptors, the negative and cognitive symptoms are the result of reduced dopamine acting on D1 receptors in the prefrontal cortex [44].

8.4.2.2 Neurodevelopmental Hypothesis

As mentioned, some researchers find that the dopaminergic hypothesis may not fully explain schizophrenia. Another hypothesis that has garnered support is the neurodevelopmental hypothesis. This is the proposition that schizophrenia, as a disorder that presents in early adulthood, might actually find its origin much earlier when the brain is developing [45]. Specifically, some studies suggest that disruption of brain development may occur as ealier as late first trimester or early second trimester [47]. This disruption may be the result of genetic and/or environmental factors that cause neurodevelopmental difficulties [48]. Some of the perinatal and obstetric environmental factors put forth as potential causes of dysfunctional neurodevelopment in schizophrenia include periventricular hemorrhages, hypoxia, ischemia, viral infection, influenza exposure [49, 50, 51, 52, 53, 54]. These genetic and environmental factors affect neurodevelopment in schizophrenia by specifically impacting "early migration of neurons and glia, cell proliferation, axonal outgrowth, synaptogenesis, and apoptosis" [46].

This idea is supported by the fact that schizophrenia commonly co-occurs with other neurodevelopmental disorders such as autism spectrum disorder, intellectual disability, and attention-deficit hyperactivity disorder pointing to a potentially shared pathway [45].

8.4.2.3 Synaptic Hypothesis

In recent years, the synaptic hypothesis has gained more momentum. Dr. Irwin Feinberg was the first to propose in 1982 that schizophrenia may be the result of faulty synatipc pruning [55]. Then the idea was further developed and suggests that schizophrenia is actually caused by excessive cortical synaptic pruning in the prefrontal cortex as well as a lack of pruning of subcortical synpases [56]. Current literature proposes additional improvements to the synaptic hypothesis in the form of a "multi-hit" model [57]. This model suggests that certain genetic risk factors may make an individual more vulnerable to synaptic elimination. Then in the presence of a stressor in the late neurodevelopmental period, dysfunction synaptic pruning via glial activity may occur. This synaptic elimination then causes pyramidal neuron dysfunction in the cortex leading to negative and cognitive symptoms. Additionally, synaptic elimination leads to the disinhibition of the mesostriatal regions and excessive dopamine activity thus causing positive symptoms of psychosis [57].

8.5 Risk Factors

As mentioned above, there is still a great deal that is unknown regarding the pathology and eitology of schizophrenia. As a result, there has been a push in current literature to understand the risk factors associated with schizophrenia in the hopes of illuminating mechanisms of the disorder [72].

8.5.1 Environmental Factors

There are many environmental/non-genetic factors that have been shown to be associated with schizophrenia. The presence of certain socioenvironmental risk factors during early childhood and adolescence have been linked to early onset schizophrenia [72]. These factors include childhood trauma, migration and minority status, traumatic brain injury and substance use [73, 74, 75, 76, 77, 78]. Additional studies showed that birth complications including fetal hypoxia, preclampsia, malformations and vacuum extraction are associated with schizophrenia development in the child [80, 81]. Current data also suggests that maternal stress during the first trimester may be linked to schizophrenia [82]. Infections, especially influenza virus and measles, during pregnancy is also a risk factor for schizophrenia [83, 84]. For late onset schizophrenia, stressors later in life outside of the timeframe of key developmental periods and feelings of loss have been identified as potential risk factors [79].

8.5.2 Genetic Factors

Current literature has shown that there are many genetic factors associated with schizophrenia. Certain studies suggest that the heritability of schizophrenia may be up to 80% [89]. Early onset schizophrenia is more common in males than females [91]. Another genetic factor linked to schizophrenia development is a positive family history of schziophrenia and other affective disorders [94, 95]. In comparison, females seem to be at a higher risk for developing late onset schizophrenia [90]. Additionally, a family history of affective disorders and cluster A personality traits are frequently associated with late onset schizophrenia [92, 93].

Mutations of several gene families have been implicated in the development of schizophrenia. Genes associated with the functions of oligodendrocyte and myelin have been linked to schizophrenia development [87]. When it comes to the identification of a specific gene associated with the development of schizophrenia, many have been identified. However, none have been shown to have a genetic association with schizophrenia in every replicated study [88]. Currently, the genes suggested to be risk factors in the development of schizophrenia are [46]:

- Neuregulin (*NRG1*)
- Dysbindin (*DTNBP1*)
- G72 (*G72*)
- D-amino acid oxidase (*DAOO*)
- RGS4 (*RGS4*)
- Catechol-*O*-methyltransferase (*COMT*)
- Proline dehydrogenase (*PRODH*)
- Reelin (*RELN*)

8.5.3 Comorbidities

There are certain disorders that tend to co-occur with schizophrenia and often precede schziophrenia onset. Psychiatric conditions including depression and anxiety as well as substance use disorder are some of the most common comorbidities [85, 86].

8.6 Future Directions

Schizophrenia is an extremely heterogenous disorder as evidence of the DSM-5 distinction of schizophrenia as a disorder that exists on a spectrum. As a result, there are many pathologies, etiologies and risk factors that have been identified. Further research to determine the mechanisms that underlie this disorder are paramount in order to best care for patients. Additionally, there has been data that suggests gender difference in schizophrenia prevalence. It is important to further illuminate how schizophrenia presents in all genders. In addition, data on the prevalence and presentation of schizophrenia in individuals that do not identify within the gender binary is needed.

References

[1] Luvsannyam, E., M. S. Jain, M. K. L. Pormento, H. Siddiqui, A. R. A. Balagtas, B. O. Emuze et al. 2022. Neurobiology of schizophrenia: a comprehensive review. Cureus 14(4).

[2] Engemann, K., C. B. Pedersen, E. Agerbo, L. Arge, A. D. Børglum, C. Erikstrup et al. 2020. Association between childhood green space, genetic liability, and the incidence of schizophrenia. Schizophrenia Bulletin 46(6): 1629–1637.

[3] Hjorthøj, C., A. E. Stürup, J. J. McGrath and M. Nordentoft. 2017. Years of potential life lost and life expectancy in schizophrenia: a systematic review and meta-analysis. The Lancet Psychiatry 4(4): 295–301.
[4] Charlson, F. J., A. J. Ferrari, D. F. Santomauro, S. Diminic, E. Stockings, J. G. Scott et al. 2018. Global epidemiology and burden of schizophrenia: findings from the global burden of disease study 2016. Schizophrenia Bulletin 44(6): 1195–1203.
[5] Desai, P. R., K. A. Lawson, J. C. Barner and K. L. Rascati. 2013. Estimating the direct and indirect costs for community-dwelling patients with schizophrenia. Journal of Pharmaceutical Health Services Research 4(4): 187–194.
[6] Hakulinen, C., J. J. McGrath, A. Timmerman, N. Skipper, P. B. Mortensen, C. B. Pedersen et al. 2019. The association between early-onset schizophrenia with employment, income, education, and cohabitation status: nationwide study with 35 years of follow-up. Social Psychiatry and Psychiatric Epidemiology 54: 1343–1351.
[7] Chen, L., A. Selvendra, A. Stewart and D. Castle. 2018. Risk factors in early and late onset schizophrenia. Comprehensive Psychiatry 80: 155–162.
[8] Werry, J. S. 1992. Child and adolescent (early onset) schizophrenia: a review in light of DSM-III-R. Journal of Autism and Developmental Disorders 22(4): 601–624.
[9] Remschmidt, H. E., E. Schulz, M. Martin, A. Warnke and G. E. Trott. 1994. Childhood-onset schizophrenia: history of the concept and recent studies. Schizophrenia Bulletin 20(4): 727–745.
[10] Rammou, A., H. L. Fisher, S. Johnson, B. Major, N. Rahaman, N. Chamberlain-Kent et al. 2019. Negative symptoms in first-episode psychosis: Clinical correlates and 1-year follow-up outcomes in London Early Intervention Services. Early Intervention in Psychiatry 13(3): 443–452.
[11] Luoma, S., H. Hakko, T. Ollinen, M. R. Järvelin and S. Lindeman. 2008. Association between age at onset and clinical features of schizophrenia: the Northern Finland 1966 birth cohort study. European Psychiatry 23(5): 331–335.
[12] Hollis, C. 2003. Developmental precursors of child-and adolescent-onset schizophrenia and affective psychoses: diagnostic specificity and continuity with symptom dimensions. The British Journal of Psychiatry 182(1): 37–44.
[13] Coulon, N. 2015. Liens entre troubles du spectre autistique et schizophrénies précoces? (Doctoral dissertation, Université Pierre et Marie Curie-Paris VI).
[14] Joa, I., J. O. Johannessen, J. Langeveld, S. Friis, I. Melle, S. Opjordsmoen et al. 2009. Baseline profiles of adolescent vs. adult-onset first-episode psychosis in an early detection program. Acta Psychiatrica Scandinavica 119(6): 494–500.
[15] Schimmelmann, B. G., P. Conus, S. Cotton, P. D. McGorry and M. Lambert. 2007. Pre-treatment, baseline, and outcome differences between early-onset and adult-onset psychosis in an epidemiological cohort of 636 first-episode patients. Schizophrenia Research 95(1-3): 1–8.
[16] Coulon, N., O. Godin, E. Bulzacka, C. Dubertret, J. Mallet, G. Fond et al. 2020. Early and very early-onset schizophrenia compared with adult-onset schizophrenia: French FACE-SZ database. Brain and Behavior 10(2): e01495.
[17] Bhati, M. T. 2013. Defining psychosis: the evolution of DSM-5 schizophrenia spectrum disorders. Current Psychiatry Reports 15: 1–7.
[18] Tandon, R., W. Gaebel, D. M. Barch, J. Bustillo, R. E. Gur, S. Heckers et al. 2013. Definition and description of schizophrenia in the DSM-5. Schizophrenia Research 150(1): 3–10.
[19] McCutcheon, R. A., R. S. Keefe and P. K. McGuire. 2023. Cognitive impairment in schizophrenia: aetiology, pathophysiology, and treatment. Molecular Psychiatry 28(5): 1902–1918.
[20] Kay, S. R., A. Fiszbein and L. A. Opler. 1987. The positive and negative syndrome scale (PANSS) for schizophrenia. Schizophrenia Bulletin 13(2): 261–276.
[21] Andreasen, N. C. and S. Olsen. 1982. Negative v positive schizophrenia: Definition and validation. Archives of General Psychiatry 39(7): 789–794.
[22] Schlagenhauf, F. and P. Sterzer. 2023. Psychotic Disorders ("Schizophrenia"). pp. 263–282. In Psychoneuroscience. Berlin, Heidelberg: Springer Berlin Heidelberg.
[23] Canadian Agency for Drugs and Technologies in Health. Validity of Psychiatric Symptom Scales and Clinical Implications. 2011 Dec.

[24] Schiffman, J., E. Walker, M. Ekstrom, F. Schulsinger, H. Sorensen and S. Mednick. 2004. Childhood videotaped social and neuromotor precursors of schizophrenia: a prospective investigation. Am. J. Psychiatry 161(11): 2021–2027.

[25] Coyle, J. T. 2017. Schizophrenia: basic and clinical. Neurodegenerative Diseases: Pathology, Mechanisms, and Potential Therapeutic Targets 255–280.

[26] Perlini, C., M. Bellani, B. Besteher, I. Nenadić and P. Brambilla. 2018. The neural basis of hostility-related dimensions in schizophrenia. Epidemiology and Psychiatric Sciences 27(6): 546–551.

[27] Positive and Negative Syndrome Scale for Schizophrenia (PANSS). Positive and negative syndrome scale for schizophrenia (PANSS). (n.d.). https://reference.medscape.com/calculator/860/positive-and-negative-syndrome-scale-for-schizophrenia-panss#.

[28] Notaras, M., A. Lodhi, F. Dündar, P. Collier, N. M. Sayles, H. Tilgner et al. 2022. Schizophrenia is defined by cell-specific neuropathology and multiple neurodevelopmental mechanisms in patient-derived cerebral organoids. Molecular Psychiatry 27(3): 1416–1434.

[29] Harrison, P. J., D. A. Lewis and J. E. Kleinman. 2010. Neuropathology of schizophrenia. Schizophrenia 372–392.

[30] Lewis, D. A., A. A. Curley, J. R. Glausier and D. W. Volk. 2012. Cortical parvalbumin interneurons and cognitive dysfunction in schizophrenia. Trends Neurosci 35(1): 57–67.

[31] Spokes, E. G. 1979. GABA in Huntington's chorea, Parkinsonism and schizophrenia. Adv Exp Med Biol. 123: 461–473.

[32] Carlén, M., K. Meletis, J. H. Siegle, J. A. Cardin, K. Futai, D. Vierling-Claassen et al. 2012. A critical role for NMDA receptors in parvalbumin interneurons for gamma rhythm induction and behavior. Mol Psychiatry 17(5): 537–548.

[33] Pierri, J. N., C. L. Volk, S. Auh, A. Sampson and D. A. Lewis. 2001. Decreased somal size of deep layer 3 pyramidal neurons in the prefrontal cortex of subjects with schizophrenia. Arch Gen Psychiatry 58(5): 466–473.

[34] Rosoklija, G., G. Toomayan, S. P. Ellis, J. Keilp, J. J. Mann, N. Latov et al. 2000. Structural abnormalities of subicular dendrites in subjects with schizophrenia and mood disorders: preliminary findings. Arch Gen Psychiatry 57: 349–356.

[35] Sweet, R. A., R. A. Henteleff, W. Zhang, A. R. Sampson and D. A. Lewis. 2009. Reduced dendritic spine density in auditory cortex of subjects with schizophrenia. Neuropsychopharmacology 34(3): 74–89.

[36] Haug, J. O. 1962. Pneumoencephalographic studies in mental disease. Acta Psychiatr Scand Suppl 38: 1–104.

[37] Van Horn, J. D. and I. C. McManus. 1992. Ventricular enlargement in schizophrenia. A meta-analysis of studies of the ventricle: brain ratio (VBR). Br J. Psychiatry 160: 687–697.

[38] Kuperberg, G. R., M. R. Broome, P. K. McGuire, A. S. David, M. Eddy, F. Ozawa et al. 2003. Regionally localized thinning of the cerebral cortex in schizophrenia. Arch Gen Psychiatry 60(9): 878–888.

[39] Konopaske, G. T., N. Lange, J. T. Coyle and F. M. Benes. 2014. Prefrontal cortical dendritic spine pathology in schizophrenia and bipolar disorder. JAMA Psychiat 71: 1323–1331.

[40] Ho, B. C., N. C. Andreasen, S. Ziebell, R. Pierson and V. Magnotta. 2011. Long-term antipsychotic treatment and brain volumes: a longitudinal study of first-episode schizophrenia. Arch Gen Psychiatry 68(2): 128–137.

[41] Sahito, A. M., M. M. Hasan, S. Shaikh, A. F. Sahito, S. A. M. Shah, M. A. Butt et al. 2023. Different Phases of Schizophrenia Patients: From the psychological perspective. pp. 197–213. In Cognizance of Schizophrenia: A Profound Insight into the Psyche. Singapore: Springer Nature Singapore.

[42] Carlsson, A. 1988. The current status of the dopamine hypothesis of schizophrenia. Neuropsychopharmacology: official publication of the American College of Neuropsychopharmacology 1(3): 179–186.

[43] Snyder, S. H. 1976. The dopamine hypothesis of schizophrenia: focus on the dopamine receptor. Am J. Psychiatry 133(2): 197–202.

[44] Toda, M. and A. Abi-Dargham. 2007. Dopamine hypothesis of schizophrenia: making sense of it all. Current Psychiatry Reports 9(4): 329–336.

[45] Owen, M. J., M. C. O'Donovan, A. Thapar and N. Craddock. 2011. Neurodevelopmental hypothesis of schizophrenia. The British Journal of Psychiatry 198(3): 173–175.
[46] Fatemi, S. H. and T. D. Folsom. 2009. The neurodevelopmental hypothesis of schizophrenia, revisited. Schizophrenia Bulletin 35(3): 528–548.
[47] Fatemi, S. H. 2005. Neuropsychiatric Disorders and Infection. CRC Press.
[48] Rapoport, J. L., A. M. Addington, S. Frangou and M. R. C. Psych. 2005. The neurodevelopmental model of schizophrenia: update 2005. Molecular Psychiatry 10(5): 434–449.
[49] Ebert, M. H., P. T. Loosen and B. Nurcombe. 2000. Current diagnosis and treatment in psychiatry.
[50] Lieberman, J. A., T. Stroup and D. O. Perkins. 2006. The American Psychiatric Publishing Textbook of Schizophrenia. American Psychiatric Publishing, Inc.
[51] Lewis, D. A. 2001. Retroviruses and the pathogenesis of schizophrenia. Proceedings of the National Academy of Sciences 98(8): 4293–4294.
[52] Karlsson, H., S. Bachmann, J. Schröder, J. McArthur, E. F. Torrey and R. H. Yolken. 2001. Retroviral RNA identified in the cerebrospinal fluids and brains of individuals with schizophrenia. Proceedings of the National Academy of Sciences 98(8): 4634–4639.
[53] Hare, E. H., J. S. Price and E. Slater. 1972. Schizophrenia and season of birth. The British Journal of Psychiatry 120(554): 124–125.
[54] Machón, R. A., S. A. Mednick and F. Schulsinger. 1983. The interaction of seasonality, place of birth, genetic risk and subsequent schizophrenia in a high risk sample. The British Journal of Psychiatry 143(4): 383–388.
[55] Feinberg, I. 1982. Schizophrenia: caused by a fault in programmed synaptic elimination during adolescence?. Journal of Psychiatric Research 17(4): 319–334.
[56] Keshavan, M. S., S. Anderson and J. W. Pettergrew. 1994. Is schizophrenia due to excessive synaptic pruning in the prefrontal cortex? The Feinberg hypothesis revisited. Journal of Psychiatric Research 28(3): 239–265.
[57] Howes, O. D. and E. C. Onwordi. 2023. The synaptic hypothesis of schizophrenia version III: a master mechanism. Molecular Psychiatry 28(5): 1843–1856.
[58] McGorry, P. D., B. Nelson, S. Goldstone and A. R. Yung. 2010. Clinical staging: a heuristic and practical strategy for new research and better health and social outcomes for psychotic and related mood disorders. The Canadian Journal of Psychiatry 55(8): 486–497.
[59] Addington, J. 2003. The prodromal stage of psychotic illness: observation, detection or intervention?. Journal of Psychiatry and Neuroscience 28(2): 93–97.
[60] Cornblatt, B., T. Lencz and M. Obuchowski. 2002. The schizophrenia prodrome: treatment and high-risk perspectives. Schizophrenia Research 54(1-2): 177–186.
[61] Hoffman, B. F. 1982. The stages of schizophrenia and their management. Canadian Family Physician Medecin de famille canadien 28: 2046–2050.
[62] Roberts, J., A. Gracia Canales, S. Blanthorn-Hazell, A. Craciun Boldeanu and D. Judge. 2018. Characterizing the experience of agitation in patients with bipolar disorder and schizophrenia. BMC Psychiatry 18: 1–8.
[63] McGlashan, T. H. 1988. A selective review of recent North American long-term followup studies of schizophrenia. Schizophrenia Bulletin 14(4): 515–542.
[64] Shepherd, M., D. Watt, I. Falloon and N. Smeeton. 1989. The natural history of schizophrenia: a five-year follow-up study of outcome and prediction in a representative sample of schizophrenics. Psychological Medicine Monograph Supplement 15: 1–46.
[65] Andreasen, N. C., W. T. Carpenter Jr, J. M. Kane, R. A. Lasser, S. R. Marder and D. R. Weinberger. 2005. Remission in schizophrenia: proposed criteria and rationale for consensus. American Journal of Psychiatry 162(3): 441–449.
[66] Correll, C. U., T. Kishimoto, J. Nielsen and J. M. Kane. 2011. Quantifying clinical relevance in the treatment of schizophrenia. Clinical Therapeutics 33(12): B16–B39.
[67] Lee, B. J., S. W. Kim, J. J. Kim, J. C. Yu, K. Y. Lee, S. H. Won et al. 2020. Defining treatment response, remission, relapse, and recovery in first-episode psychosis: a survey among Korean experts. Psychiatry Investigation 17(2): 163.
[68] Novick, D., W. Montgomery, Y. Cheng, V. Moneta and J. M. Haro. 2015. Impact of negative symptoms on quality of life in patients with schizophrenia. Value in Health 18(7): A836–A837.

[69] Fountoulakis, K. N., E. Dragioti, A. T. Theofilidis, T. Wikilund, X. Atmatzidis, I. Nimatoudis et al. 2019. Staging of schizophrenia with the use of PANSS: an international multi-center study. International Journal of Neuropsychopharmacology 22(11): 681–697.

[70] Agius, M., C. Goh, S. Ulhaq and P. McGorry. 2010. The staging model in schizophrenia, and its clinical implications. Psychiatria Danubina 22(2): 211–220.

[71] Roy, A. 1982. Suicide in chronic schizophrenia. The British Journal of Psychiatry 141(2): 171–177.

[72] Chen, L., A. Selvendra, A. Stewart and D. Castle. 2018. Risk factors in early and late onset schizophrenia. Comprehensive Psychiatry 80: 155–162.

[73] Van Os, J., G. Kenis and B. P. Rutten. 2010. The environment and schizophrenia. Nature 468(7321): 203–212.

[74] Bendall, S., H. J. Jackson, C. A. Hulbert and P. D. McGorry. 2008. Childhood trauma and psychotic disorders: a systematic, critical review of the evidence. Schizophrenia Bulletin 34(3): 568–579.

[75] Bourque, F., E. van der Ven and A. Malla. 2011. A meta-analysis of the risk for psychotic disorders among first-and second-generation immigrants. Psychological Medicine 41(5): 897–910.

[76] Orlovska, S., M. S. Pedersen, M. E. Benros, P. B. Mortensen, E. Agerbo and M. Nordentoft. 2014. Head injury as risk factor for psychiatric disorders: a nationwide register-based follow-up study of 113,906 persons with head injury. American Journal of Psychiatry 171(4): 463–469.

[77] Moore, T. H., S. Zammit, A. Lingford-Hughes, T. R. Barnes, P. B. Jones, M. Burke et al. 2007. Cannabis use and risk of psychotic or affective mental health outcomes: a systematic review. The Lancet 370(9584): 319–328.

[78] Barkus, E. and R. M. Murray. 2010. Substance use in adolescence and psychosis: clarifying the relationship. Annual Review of Clinical Psychology 6(1): 365–389.

[79] Howard, R., P. V. Rabins, M. V. Seeman, D. V. Jeste and T. I. Late-Onset. 2000. Late-onset schizophrenia and very-late-onset schizophrenia-like psychosis: an international consensus. American Journal of Psychiatry 157(2): 172–178.

[80] Boydell, J. 2001. Risk factors for schizophrenia. Expert Review of Neurotherapeutics 1(2): 183–191.

[81] Vigod, S. N., P. A. Kurdyak, C. L. Dennis, A. Gruneir, A. Newman, M. V. Seeman et al. 2014. Maternal and newborn outcomes among women with schizophrenia: a retrospective population-based cohort study. BJOG: An International Journal of Obstetrics and Gynaecology 121(5): 566–574.

[82] Janoutová, J., P. Janácková, O. Sery, T. Zeman, P. Ambroz, M. Kovalová et al. 2016. Epidemiology and risk factors of schizophrenia. Neuroendocrinology Letters 37(1): 1–8.

[83] Landreau, F., P. Galeano, L. R. Caltana, L. Masciotra, A. Chertcoff, A. Pontoriero et al. 2012. Effects of two commonly found strains of influenza A virus on developing dopaminergic neurons, in relation to the pathophysiology of schizophrenia. PloS One 7(12): e51068.

[84] Yolken, R. 2004. Viruses and schizophrenia: a focus on herpes simplex virus. Herpes: The Journal of the IHMF 11: 83A–88A.

[85] Buckley, P. F., B. J. Miller, D. S. Lehrer and D. J. Castle. 2009. Psychiatric comorbidities and schizophrenia. Schizophrenia Bulletin 35(2): 383–402.

[86] Achim, A. M., M. Maziade, É. Raymond, D. Olivier, C. Mérette and M. A. Roy. 2011. How prevalent are anxiety disorders in schizophrenia? A meta-analysis and critical review on a significant association. Schizophrenia Bulletin 37(4): 811–821.

[87] Hakak, Y., J. R. Walker, C. Li, W. H. Wong, K. L. Davis, J. D. Buxbaum et al. 2001. Genome-wide expression analysis reveals dysregulation of myelination-related genes in chronic schizophrenia. Proceedings of the National Academy of Sciences 98(8): 4746–4751.

[88] Levitt, P., P. Ebert, K. Mirnics, V. L. Nimgaonkar and D. A. Lewis. 2006. Making the case for a candidate vulnerability gene in schizophrenia: convergent evidence for regulator of G-protein signaling 4 (RGS4). Biological Psychiatry 60(6): 534–537.

[89] Hosak, L. 2013. New findings in the genetics of schizophrenia. World Journal of Psychiatry 3(3): 57.

[90] Harris, M. J. and D. V. Jeste. 1988. Late-onset schizophrenia: an overview. Schizophrenia Bulletin 14(1): 39–55.

[91] Ochoa, S., J. Usall, J. Cobo, X. Labad and J. Kulkarni. 2012. Gender differences in schizophrenia and first-episode psychosis: A comprehensive literature review. Schizophrenia Research and Treatment 2012(1): 916198.
[92] Howard, R. J., C. Graham, P. Sham, J. Dennehey, D. J. Castle, R. Levy et al. 1997. A controlled family study of late-onset non-affective psychosis (late paraphrenia). The British Journal of Psychiatry 170(6): 511–514.
[93] Wynn Owen, P. A. and D. J. Castle. 1999. Late-onset schizophrenia: epidemiology, diagnosis, management and outcomes. Drugs and Aging 15: 81–89.
[94] Maier, W., D. Lichtermann, P. Franke, R. Heun, P. Falkai and M. Rietschel. 2002. The dichotomy of schizophrenia and affective disorders in extended pedigrees. Schizophrenia Research 57(2-3): 259–266.
[95] Faridi, K., N. Pawliuk, S. King, R. Joober and A. K. Malla. 2009. Prevalence of psychotic and non-psychotic disorders in relatives of patients with a first episode psychosis. Schizophrenia Research 114(1-3): 57–63.
[96] Adityanjee, N., Y. A. Aderibigbe, D. Theodoridis and W. V. R. Vieweg. 1999. Dementia praecox to schizophrenia: the first 100 years. Psychiatry and Clinical Neurosciences 53(4): 437–448.

Chapter 9

Psychological Mechanisms of Addiction

Nicholas B. Wells,[1,*] *Sophia Marra,*[1] *Emre Tokgoz*[2] and *Percin Karakol*[3]

9.1 Overview of the Brain's Reward System

Addiction takes hold by changing the way our brain rewards us for positive behavior. The primary center for reward in the brain lies in the midbrain, specifically within the ventral tegmental area, or VTA. Another term for this reward region is the mesolimbic system. Projections from the VTA to other areas of the midbrain shown above such as the dorsal striatum, prefrontal cortes, hippocampus, nucleus accumbens and amygdala all play a role in the processing of and guidance toward rewarding behavior. An external factor such as a drug or a game that tells us that it is a positive experience for us, our brain will guide us back to that behavior and the positive stimulus that comes with it; Repeated exposure can result in a desire to acquire the need for such activity as the brain continues to seek out the stimulus while in many cases it can be resistant to it [1, 2].

By dispersing the neurotransmitter Dopamine, the VTA influences various other areas in the midbrain in response to a behavior or substance. The VTA within the midbrain is responsible for many activities such as movement, thinking, reward, motivation, and even reproductive tendencies. While the downstream impacts of dopamine on behavior are well known, much about how the VTA functions remains difficult to pin down. Figure 1 below outlines many targets of dopamine in the midbrain and beyond [3, 4].

[1] Frank H. Netter School of Medicine, Quinnipiac University, North Haven, CT, 06473, USA.
[2] School of Engineering and Technology, Farmingdale State College, New York, 11735 USA.
[3] Başakşehir Cam and Sakura City Hospital Plastic Reconstructive and Aesthetic Surgery, Istanbul, Turkiye.
* Corresponding author: Nicholas.Wells@quinnipiac.edu

DOI: 10.1201/9781003646044-9

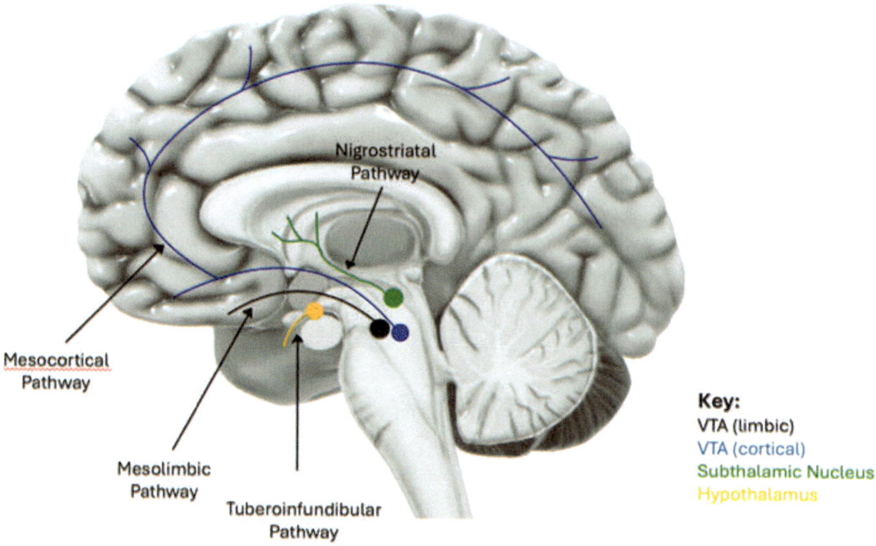

Fig. 1. Dopamine impacting regions in the brain.

Two regions specifically related to addiction behavioral patterns include the dorsal striatum (DTS) and hippocampus (HIP). The dorsal striatum aids in habit formation. Dopaminergic activation of the dorsal striatum occurs via D1 and D2 GABA-producing medium spiny neurons and cholinergic interneurons [5].

The medium spiny neurons are more involved in emotional regulation as well as motor function and learning while the cholinergic neurons, while much scarcer, are more closely related to learning in the context of reward. The association of cholinergic neurons with motor learning and reward provides a link between the dorsal striatum and addictive patterns of behavior [6].

9.1.1 The Direct and Indirect Dopamine Pathways

The more numerous dopaminergic neurons, the GABA-producing medium spiny neurons, form two pathways: direct (D1) and indirect (D2) within the basal ganglia, which receives input from the dorsal striatum. The specific function of each pathway (Kim) can be categorized by the impact each pathway has on movement. The direct pathway inhibits an area called the substantia nigra pars reticulata, which in turn releases the thalamus from inhibition, generating movement. The indirect pathway provokes the opposite response by stimulating the subthalamic nucleus and subsequently increasing activity of the substantia nigra pars reticulata, inhibiting movement through increased inhibition of the thalamus. The pathways act in opposition, with one initiating movement and the other limiting movement. Neuronal loss in the substantia nigra pars compacta, which normally inhibits the indirect pathway, can lead to pathologies such as Parkinson's disease [7, 8].

The hippocampus forms addictive behavior through memory. While not responsible for creating the physical behaviors associated with addiction, the hippocampus works with the amygdala to use memory to guide behavior. By creating a link between memories and past reward, the hippocampus helps us recognize environmental cues that will likely lead to reward. In doing so, the hippocampus and amygdala can establish associations that drive people toward reward-seeking behavior, in many cases manifesting in addictive patterns of behavior.

Possibly the most important piece of the mesolimbic system is the Nucleus Accumbens (NA). A component of the ventral striatum, the NA helps to process reward and incentive-based behavior. By receiving dopaminergic stimulation via the VTA, the NA appraises rewarding behavior, giving value to a positive stimulus that will influence behavior in the future. An addiction to a substance or behavior will prioritize seeking out that stimulus based on how high the perceived value of that stimulus is. The Nucleus Accumbens binds a reward to its appraisal value, creating an incentive [9].

The various pathways associated with dopaminergic stimulation within the mesolimbic system all help to establish patterns of behavior that seek reward. Excessive stimulation or overstimulation of these pathways can lead to a greater propensity toward reward seeking. While some drugs directly manipulate release of dopamine, other stimuli such as sexual activity and gambling have a more indirect effect while creating similar patterns of behavior. The mesolimbic system plays a large role in gene expression related to reward-seeking behaviors (food consumption, drug use, sexual activity, social interaction). When a reward is repeatedly encountered, the mesolimbic system establishes a feedback loop to not only acquire more of the stimulus but prepare the brain via gene expression for more of the stimulus [10].

9.2 Categorizing Drug Addiction as a Disease

While originally described as a bad habit, addiction is now thought to be more closely aligned with a disease state. Habit formation is certainly a cornerstone of addiction, but relapse rates and physical symptoms of withdrawal following abstinence are more indicative of a chronic disease. Relapse rates among substance abusers increase as more time passes, just as survival rates decrease significantly over time. Substance abuse comes with the negative effect of altering the previously discussed neurons in the midbrain. Neurons will react more strongly to an abused substance while simultaneously decreasing the response to environmental cues not related to the substance in question. The neuron will also demonstrate a heightened response to negative stimuli, creating a behavioral pattern that favors increased use of a substance and aversion to the negative effects of not having the substance. Delaying a reward linked to a substance creates what drug abusers might describe as a craving, which is the brain's way of establishing a desire, or more accurately a need, to have the substance again. The delay in acquiring the reward that the brain has been accustomed to through habit formation is what drives the behavioral pattern known as relapsing, wherein the user will return to the substance or stimulus after an unusually long hiatus. This shift from a positive reinforcement pattern, in which the abuser is seeking a reward, to a negative reinforcement pattern of avoiding withdrawal symptoms, is a hallmark of addiction [11].

9.3 Genetic and Environmental Components of Addiction

A highly debated discussion point in addiction medicine is the comparison of genetic versus environmental factors. Identifying risk factors for addiction is helpful for evaluating current addictive states and preventing future addictive habits from forming. While the dopamine reward pathway in the brain is responsible primarily for creating an addiction, many other factors contribute to the disease state. Various genes have been identified that are associated with addiction.

Substance abuse and the ensuing addiction can be traced to genes that share similar characteristics. Genes directly involved in stress reactivity can establish a predisposition toward reward seeking during times of stress. Other genes that code for pathways involved with impulsivity or pleasure seeking can also lead to addiction. The way that these genes are structured also matters from an epigenetic perspective; they can be predisposed to cellular changes through activation and silencing. Factors that contribute to neural remodeling and drug abuse include genetics, stressors, developmental changes, physiology, and drug pharmacokinetics. When any number of these factors combine, a certain phenotype can arise that drives an individual toward drug abuse and eventual addiction [12].

Dopamine and the reward system remain at the crux of the addiction disease, but the pathway itself is subject to not just overstimulation but also genetic variation. Lasting change in the way our brains process reward because of substance abuse. The opioid pathway determines the brain's response to a variety of substances, primarily heroin, suggesting that predisposition to many substances could be a result of genetic alterations to the opioid pathway. Alcohol, which will be discussed further in a later section, generates a reward response though GABA neurons, which are responsible for reward responses for other substances such as nicotine. The inverse is true regarding nicotine receptors and alcohol, indicating that modification of pathways at the genetic level, due to inherited genes or abuse of one substance, can increase the addictive potential of another substance [13].

9.3.1 Epigenetics

While changes in the actual sequence of DNA in genes can impact addictive behavior, another mechanism of addiction is epigenetic change. A gene can respond differently to changes in regulatory proteins and DNA expression. Through reward-inducing substances, change can be induced in the genes coding for those reward pathways. Change can occur at the level of protein structure or even signal transduction, affecting the way the reward for a substance is interpreted as the substance is used in repetition. Epigenetic changes induced by substance abuse can last for years and continue to have downstream impacts even after abuse ends. Change can also occur through neuroendocrine modification and environmental factors that alter the pathways of our brain. Epigenetic treatment is gaining traction to address addiction because it harnesses the plasticity and reversibility of the genes being changed [14].

9.3.2 The Triad of Addiction, Stress, and Socioeconomic Distress

Genetics is a major factor in the likelihood of substance abuse and addiction development. At the epigenetic level, even external environmental factors can contribute to change. While the environment can evoke epigenetic change, it can also have a direct effect on the propensity toward substance use and addictive behavior. A paper by Fink et al. describes a strong association between overdose deaths and reduced socioeconomic status. Negative socioeconomic factors such as poverty and poor economies have been linked to disproportionately high rates of substance use. Deaths due to overdose are higher in areas facing social challenges such as unemployment and deindustrialization. The environmental impact on addiction is two-fold: physical resources and the psychosocial consequences of resource deprivation.

Unemployment or general financial strife can directly increase substance use and addiction persistence by limiting both economic resources and access to healthcare. Limited or absent income results not only in an increase in stress but also a lack of physical resources like healthy food and transportation. Poverty also limits baseline access to healthcare, including addiction treatment, as low-income areas are much less likely to have access to quality healthcare [16].

Drug use involves several progressive stages: opportunity, initiation, regular use, dependence. It is proposed that specific combinations of genetic and environmental factors can drag a user along the subsequent stages of drug abuse. The jump from opportunity to initiation can be due to the presence of genes associated with novelty seeking and/or risk taking (genetic) combined with a lack of parental monitoring (environmental). Similarly, initial use can quickly progress to regular or heavy use when genes tasked with subjective appraisal of effect are present alongside increased substance availability. Abuse and dependence can be more likely in a person with altered drug metabolism genes and a history of peer abuse or childhood trauma. Many of these environmental factors correlate directly with financial strife [15].

Piggybacking on financial challenges is the psychosocial impact of stress. When discussing the relationship between stress, health, and addiction, it is important to note that each factor positively influences the other. Increased stress breeds negative health impacts and higher rates of substance use and addiction. Substance use decreases overall health and increases stress with chronic use. Poor health or limited access to healthcare reduces overall health while increasing the likelihood of substance abuse and addiction. With this relationship in mind, the health consequences of stress can therefore have a direct correlation with addiction, and vice versa.

9.3.3 Direct Effects of Stress on Health

The physiological response to stress, colloquially known as the fight-or-flight response, can be dangerous in states of chronic stimulation, and can also lead to increased rates of substance abuse. While helpful for addressing a transient threat, unrelenting issues provoking a stress response can increase the risk of chronic disease and decrease immunity. The largest variables in determining the negative effects of stress on health are frequency, severity, and duration of perceived threats or traumatic events. Coupled with genetic predisposition and physiological characteristics, stress can lead to disease states such as high blood pressure, reduced immunity, and reduced discrimination of potentially stressful stimuli [17].

Lack of social bonding in a study performed with young rats led to a hyperactive HPA axis and an overstimulated sympathetic nervous system [18, 19]. Deprivation of nurturing led to a more easily excitable stress response. In chronically stressed humans, increases in blood pressure due to sympathetic stimulation have long term pathological consequences such as vascular hypertrophy and hypertension. Vasculature damage from increased pressure is more likely to result in obstruction and clotting. During short-term stress, immunity suffers from disruption of cellular immunity in favor of promoting humoral immunity. The inverse occurs during chronic stress, where a prolonged exposure to stressful stimuli favors cellular immunity while sacrificing humoral immunity [20]. While cellular immunity is critical, both mechanisms of defense are necessary for good health. The effects of chronic stress are particularly malignant in elderly populations due to immunosenescence, greatly increasing the risk of viral infection by reducing the antibody response. As with the elderly population, stress in adolescence can significantly increase the likelihood of developing both mood disorders and substance use disorders, further solidifying the connection between stress and negative health outcomes [21].

The direct relationship between stress and health translates to the third part of the triad: substance abuse. Particular vulnerabilities that arise from stress are drug abuse and relapse. At the highest level, this relationship is driven by similarities between the reward pathway and stress circuitry. Beyond the overlapping circuits lies the tendency of drug abuse to result in physiological changes to the stress response system in addition to the previously discussed changes to the reward pathway. This multi-faceted effect of drug abuse takes advantage of the similarities in stress perception and reward seeking and creates an association between stress and lack of reward. Once this relationship is established, stress can trigger an increase in drug cravings, leading to stressful events more likely resulting in relapse. Overcoming substance abuse addiction then must involve psychological and/or pharmaceutical treatment of the stress response, as well as the introduction of health coping mechanisms. The two-way positive feedback loop of addiction and stress is formidable and gives insight into some of the environmental determinants of substance abuse [22].

9.4 Reinforcement and the Economon Model

9.4.1 Experience-based Decision Making and Reinforcement

There are several models for analyzing decision-making. Though multifaceted, the primary driver of our decisions as a species is reward versus punishment. In users of methamphetamine, studies have postulated that there is an association between deficient decision-making capabilities and methamphetamine abuse. Specifically, the deficit lies within the abuser's ability to accurately perceive the value of an outcome, leading to a decision that fails to maximize reward and minimize punishment. A study on methamphetamine use in Iran describes several methods of decision making, one of which is known as experience-based decision making, or EDM. In this study, methamphetamine abusers were more likely to display abnormal decision making compared to healthy controls. The abuser brain failed to properly follow experience-based decision making, leading to non-optimal choices. Specifically, abusers were found to be less susceptible to the effect of the actual reward and showed longer response times, indicating a failure to quickly and adequately appraise a decision. The abusers also had a lower overall threshold for performing an action, indicating a tendency toward impulsiveness. Thus, the abuser struggles to properly learn from past consequences and apply them to future decisions regarding reward-seeking behavior. Failure to minimize punishment for a rewarding stimulus is characteristic of methamphetamine users supported by other studies [23]. Deficiencies in learning because of substance use also increase the likelihood of relapse.

The literature also discusses a separate decision model, the reinforcement learning diffusion decision model, or RL-DDM, introduced by Pedersen et al. 2017. It combines reinforcement learning, which is based on learning through unexpected outcomes. Methamphetamine users in the Iranian study were adversely affected by a higher sensitivity to reward changes but a lower sensitivity to the reward itself, indicating an inability to make the optimal decision. Drug abusers carry a deficiency in proper decision-making that persists beyond the use of the substance and can have lasting implications in recovery and relapse rates [24].

9.4.2 Reinforcement Schedules

Humans learn behavior through reinforcement, be it positive or negative. Reinforcement on a continuous basis is the best way to teach something, which entails giving a reward each time the desired behavior is performed. The cadence at which rewards are dispensed determines the strength of the reinforcement when using a schedule other than continuous. The two main schedule types are ratio and interval, meaning the amount of time between rewards changes or the number of responses before a reward changes. A fixed or variable variation determines the schedule, giving four permutations of reinforcement schedules:

- Fixed interval – 5, 10, 20 minutes
- Variable interval – 5 minutes, 30 seconds, two hours
- Fixed Ratio – every 5th attempt is successful
- Variable ratio – random attempts are successful

With a fixed interval, a reward or stimulus is delivered according to a regular cadence, i.e., medication delivered every hour to a patient. A variable interval reinforcement schedule follows no logical sequence, like phone notifications. A fixed ratio reinforces a behavior after a certain number of events have occurred, like purchasing ten coffees and receiving a free one. The most powerful type of reinforcement schedule is the variable ratio, a schedule that delivers rewards after a random number of events or trials. Variable ratio reinforcement schedules offer no recognizable pattern of reward, which encourages continued participation in hopes of the next event triggering a reward. The most well-known example of a variable ratio schedule is gambling. Since games like slot machines and roulette have a completely random cadence of reward, people become enthralled by the games, continuing to participate in hopes of the next trial yielding a reward.

Substance abuse is a form of continuous reinforcement, which is the strongest method of reinforcement. Only the most pervasive partial reinforcement schedule is associated with addiction, in the example of gambling being a variable ratio schedule. With substance abuse, the reward from the drug is obtained every time the drug is consumed. Over time, as discussed previously, the brain's reward pathway can change, changing the reinforcement such that the reward is a relief from negative symptoms. In both cases, the reinforcement is continuous, encouraging the behavior to continue [25]. Response rate is an indication of effectiveness, signifying the number of new attempts at recreating the positive stimulus. Variable ratio reinforcement schedules offer the highest response rate. The unpredictable nature of a variable ratio schedule similar to that seen in gambling preys on human emotions such as anticipation. The possibility of a reward can also be reinforced in the setting of a casino, where one participant winning is broadcast by pleasing lights and sounds, encouraging other participants to continue with the hope of achieving the same outcome [26, 27].

9.4.3 The Economics of Drug Addiction

When dealing with the initial development of addiction, it is also important to consider environmental factors just as it is with continued use and relapse. The presence of certain factors can aid in creating an addiction where one would not otherwise develop, whether by working in tandem with genetic factors or simply through an overabundance of available drugs. Availability of a drug is arguably the greatest contributing factor to the development of an addiction. Reinforcement is an important part of drug addiction, as with development of any habit, but behavior can only be reinforced if the substance is readily available. Thus, the supply of a drug, whether due to geography or economic factors, will often determine the severity of an addiction.

In contrast to the agricultural beginnings of many addictive substances, modern drugs are not able to reliable be grown and consumed within a closed-loop system. Since humans cannot, in most cases, procure their own drugs, they rely on a supplier. Substances like tobacco and alcohol are supplied legally across the United States, greatly reducing barriers to consumption. Drugs that have been deemed more volatile and remain illegal require the user to take significantly more initiative when seeking a product when compared with the naturally grown drugs of previous centuries. Cooking methamphetamine requires a rudimentary lab and experience with chemistry. Even growing marijuana requires a significant investment in growing supplies and a large space. Barriers such as these historically inhibited consumption, as sustained operant behavior was required to fulfill each step involved in producing a drug. Drug abuse and distribution eventually came to more closely resemble a supply and demand model of economics, resulting in an epidemic of drug use perpetuated by positive reinforcement resulting from the activity of purchasing and selling drugs. This behavior is described in the paper An economon model of drug addiction [28, 29].

Closed loop drug consumption has been replaced by an economic model of drug distribution, creating a market that can be modeled similar to any good or service. The economon model requires a buyer and a seller to establish an exchange that is mutually beneficial to the degree that both parties are incentivized to continue. A supplier is reinforced by the buyer such that their supply of goods has a constant source of demand. A main difference between the exchange of a typical good or service and a drug is that not only does the ample supply incentivize the user, the inherent addictive properties of the drug also serve to reinforce the behavior. Repeated stimulation of the reward pathway will create an association of pleasure with the drug and establish strong environmental cues that will induce cravings [30, 31].

The general principle of an economon is mutual reinforcement. In a scenario specific to drug-addiction, the model requires a supplier like any other model but also a person with substance use disorder. Most drugs are passed down through a supply chain, following a model like the serial economa presented above. Money reinforces the supplier while the drug's effect and availability reinforces the user. Drug use also does not follow a typical supply-demand relationship. Since the effects of drugs are addictive, demand for a substance can become constant independent of supply. Thus, there is no problem of supplier surplus. Drug abusers can also fall victim to price manipulation by suppliers, like how a monopoly over a good or service can result in increased pricing (i.e., pharmaceuticals). A diabetic patient requires insulin to survive, allowing drug manufacturers to charge inexplicably high prices for a vial of insulin. Many addicts, due to severe psychological and/or physical dependence, have similar experience regarding their drug of choice. Demand for a drug can very quickly subvert the rules of economics through the mechanism of addiction.

9.5 Substance Abuse Disorder and Psychological Changes

The economon model involves a person with substance use disorder, but what characterizes such a disorder? Substance use disorder (SUD) is described by Cleveland clinic as a mental health condition involving problematic substance use that has negative impacts on their life. Just as the addictive properties of substances can vary, the severity of substance use disorder exists as a wide range. SUD can involve multiple substances and, in many cases, can have such significant health consequences that it can be fatal. Accompanying the health effects of SUD are the psychological changes associated with drug dependence, some of which have been touched on previously. Colloquially, these symptoms are described as "withdrawal" or "dope sickness".

A constellation of negative psychological symptoms accompanies substance use disorders. After prolonged use, the positive reinforcement of being high becomes reinforcement to alleviate negative symptoms brought on by abstinence. Among the most common symptoms are stress, anger, aggression, and emotional dysregulation.

The substance no longer becomes associated with a high and becomes a cure for any unpleasant feelings. The degree to which a user is experiencing such negative symptoms directly correlates to treatment outcomes and likelihood of relapse [32, 33, 34].

As discussed previously, the primary driver of addictive behavior is manipulation of the mesolimbic dopamine pathway. Alcohol is known to specifically activate this system, leading to dopamine release and subsequent development of a positive stimulus. While the pathway will react to continued abuse of ethanol, over time the pathway becomes gradually less responsive, a phenomenon more commonly known as "tolerance". Studies have revealed interactions between ethanol and many different receptors, including nicotinic, GABAminergic, glutamatergic, and opioid receptors. Ethanol was also shown to inhibit grehlin release in chronic users while abstaining alcoholics showed elevated serum grehlin levels. The multi-faceted impact of ethanol in the brain helps explain why ethanol abuse is so pervasive [35].

9.6 Identification and Treatment of Addiction

The addiction cycle often restarts after an addiction experiences relapse. Those suffering from addiction are highly prone to relapses which, as opposed to a slip, do not intend to follow through with any sort of treatment. More than 80% of addicts will relapse within 6 months. An article from Sohrabpour et al. describes a study in which 25% of patients in a prospective study on addiction relapsed within the first year [36]. Environmental and social triggers can follow an addict well after the brain stops seeking out a substance. The psychological background of an individual plays a major role in their susceptibility to relapse, including socioeconomic status, education level, social influence, self-efficacy, and behavioral problems. Literature suggests close relationships can play a major role in relapses, specifically friendships and family relationships. Such a collection of influencing factors thus requires a complex treatment approach which includes significant psychological education and intervention.

Sohrabpour et al. present using the theory of planned behavior to understand the condition of addiction. The theory of planned behavior suggests that intention is the primary driver of behavior. The theory neatly organizes beliefs into three categories: behavioral, normative, and control [37]. Attitudes about behavior, the way others perceive a behavior, and a person's belief in their ability to carry out a behavior, all funnel into the intention to perform an action. These constructs act independently of one another but converge with a single intention. The three-fold approach to addiction treatment uses the theory of planned behavior to influence the addictive brain to perform the action of recovery. While it is generally accepted that normative beliefs and behavioral beliefs support the behavior of recovery, perceived behavioral control can be severely diminished in addicts. Establishing positive control beliefs in addiction patients can be key in reaching a state of recovery by harnessing the more elusive of the three belief systems.

9.6.1 Recovery

Relief from addiction comes in the form of recovery. Complete recovery from addiction depends on a variety of factors. A review from Bunaciu et al. describes the concept of recovery capital, wherein likelihood of recovery is determined by the abundance of positive factors in one's life that promote growth and wellbeing as well as the absence of negative factors. The article emphasizes the importance of promoting personal, social, and community level recovery capital, while working with addicts to minimize exposure to recovery barriers. Like how beliefs can impact the relapse through the theory of planned behavior, a strong external influence can impact recovery through altering a belief.

A study on drug rehabilitation program retention showed that external motivation could be a positive factor in promoting recovery. The survival study compared patients who voluntarily entered treatment with those who opted for rehabilitation instead of serving time in prison. The results showed a longer retention period for patients who entered rehabilitation instead of prison, suggesting that external motivation can play a major role in keeping patients from relapsing or leaving rehab.

References

[1] Mokha, I. S. 2024. The psychology of addiction and factors contributing to substance abuse. EPRA International Journal of Multidisciplinary Research (IJMR) 10(2): 149–153.

[2] American Psychological Association. (n.d.). Substance use, abuse, and addiction. Retrieved from https://www.apa.org/topics/substance-use-abuse-addiction.

[3] Hou, G., M. Hao, J. Duan and M. H. Han. 2024. The formation and function of the VTA dopamine system. International Journal of Molecular Sciences 25(7): 3875.

[4] Jiménez-González, A., C. Gómez-Acevedo, A. Ochoa-Aguilar and A. Chavarría. 2022. The role of glia in addiction: Dopamine as a modulator of glial responses in addiction. Cellular and Molecular Neurobiology 42(7): 2109–2120.

[5] Lewis, R. G., E. Florio, D. Punzo and E. Borrelli. 2021. The Brain's Reward System in Health and Disease (pp. 57–69). Springer International Publishing.

[6] Guo, Q., D. Wang, X. He, Q. Feng, R. Lin, F. Xu et al. 2015. Whole-brain mapping of inputs to projection neurons and cholinergic interneurons in the dorsal striatum. PloS One 10(4): e0123381.

[7] Kim, S. Y. and W. Lim. 2024. Quantifying harmony between direct and indirect pathways in the basal ganglia: healthy and Parkinsonian states. Cognitive Neurodynamics 1–21.

[8] McGregor, M. M. and A. B. Nelson. 2019. Circuit mechanisms of Parkinson's disease. Neuron 101(6): 1042–1056.

[9] Fraser, K. M., B. J. Chen and P. H. Janak. 2024. Nucleus accumbens and dorsal medial striatal dopamine and neural activity are essential for action sequence performance. European Journal of Neuroscience 59(2): 220–237.

[10] Lewis, R. G., E. Florio, D. Punzo and E. Borrelli. 2021. The Brain's Reward System in Health and Disease (pp. 57–69). Springer International Publishing.

[11] Yuan, S., S. C. Jiang, Z. W. Zhang, Z. L. Li and J. Hu. 2024. Substance addiction rehabilitation drugs. Pharmaceuticals 17(5): 615.

[12] Ajonijebu, D. C., O. Abboussi, V. A. Russell, M. V. Mabandla and W. M. Daniels. 2017. Epigenetics: a link between addiction and social environment. Cellular and Molecular Life Sciences 74: 2735–2747.

[13] Volkow, N. D. and M. Muenke. 2012. The genetics of addiction. Human Genetics 131: 773–777.

[14] Koijam, A. S., K. D. Singh, B. S. Nameirakpam, R. Haobam and Y. Rajashekar. 2024. Drug addiction and treatment: An epigenetic perspective. Biomedicine and Pharmacotherapy 170: 115951.

[15] Hines, L. A., K. I. Morley, C. Mackie and M. Lynskey. 2015. Genetic and environmental interplay in adolescent substance use disorders. Current Addiction Reports 2: 122–129.
[16] Fink, D. S., J. P. Schleimer, K. M. Keyes, C. C. Branas, M. Cerdá, P. Gruenwald et al. 2024. Social and economic determinants of drug overdose deaths: a systematic review of spatial relationships. Social Psychiatry and Psychiatric Epidemiology 1–26.
[17] Kleshchova, O. and M. R. Weierich. 2022. The neurobiology of stress. pp. 17–65. *In*: Biopsychosocial Factors of Stress, and Mindfulness for Stress Reduction. Cham: Springer International Publishing.
[18] Schneiderman, N., G. Ironson and S. D. Siegel. 2005. Stress and health: psychological, behavioral, and biological determinants. Annu. Rev. Clin. Psychol. 1(1): 607–628.
[19] Ladd, C. O., R. L. Huot, P. Thrivikraman, C. B. Nemeroff, M. J. Meaney and P. M. Plotsky. 2000. Long-term behavioral and neuroendocrine adaptations to adverse early experience. Prog. Brain Res. 122: 79–101.
[20] Segerstrom, S. C. and G. E. Miller. 2004. Psychological stress and the human immune system: a meta-analysis of 30 years of inquiry. Psychol. Bull. 130: 601–630.
[21] Kirsch, D., C. M. Nemeroff and E. T. Lippard. 2020. Early life stress and substance use disorders: underlying neurobiology and pathways to adverse outcomes. Adversity and Resilience Science 1(1): 29–47.
[22] Sinha, R. 2007. The role of stress in addiction relapse. Current Psychiatry Reports 9(5): 388–395.
[23] Dean, A. C., A. M. Morales, G. Hellemann and E. D. London. 2018. Cognitive deficit in methamphetamine users relative to childhood academic performance: link to cortical thickness. Neuropsychopharmacology 43(8): 1745–1752.
[24] Pedersen, M. L., M. J. Frank and G. Biele. 2017. The drift diffusion model as the choice rule in reinforcement learning. Psychonomic Bulletin and Review 24: 1234–1251.
[25] Votaw, K. 2020. 18.5: Reinforcement schedules. Social Sci LibreTexts. https://socialsci.libretexts.org/Bookshelves/Psychology/Introductory_Psychology/General_Psychology_for_Honors_Students_(Votaw)/18%3A_Operant_Conditioning/18.05%3A_Reinforcement_Schedules#:~:text=Among%20the%20reinforcement%20schedules%2C%20variable,1).
[26] Winborn, L., D. P. Wacker, D. M. Richman, J. Asmus and D. Geier. 2002. Assessment of mand selection for functional communication training packages. Journal of Applied Behavior Analysis 35(3): 295–298.
[27] Romani, P. W. and A. M. Koerner. 2024. Comparison of fixed-ratio and fixed-interval schedules of reinforcement within the treatment of escape-reinforced problem behavior. Journal of Behavioral Education 1–20.
[28] Negus, S. S. 2024. An economon model of drug addiction. Psychopharmacology 241(3): 417–425.
[29] Cook, P. J. and M. J. Moore. 2002. The economics of alcohol abuse and alcohol-control policies. Health Affairs 21(2): 120–133.
[30] Galaj, E., E. D. Barrera and R. Ranaldi. 2020. Therapeutic efficacy of environmental enrichment for substance use disorders. Pharmacology Biochemistry and Behavior 188: 172829.
[31] Hsiung, H., K. Patel, H. Hundal, B. M. Baccouche and K. W. Tsao. 2022. Preventing substance abuse in adolescents: A review of high-impact strategies. Cureus 14(7).
[32] Koob, G. F. 2021. Drug addiction: hyperkatifeia/negative reinforcement as a framework for medications development. Pharmacological Reviews 73(1): 163–201.
[33] Murphy, S. M. and D. Polsky. 2016. Economic evaluations of opioid use disorder interventions. Pharmacoeconomics 34: 863–887.
[34] American Psychiatric Association. What Is a Substance Use Disorder (https://www.psychiatry.org/patients-families/addiction-substance-use-disorders/what-is-a-substance-use-disorder)? Last reviewed 4/2024. Accessed 9/9/2024.
[35] Söderpalm, B. and M. Ericson. 2024. Alcohol and the dopamine system. International Review of Neurobiology 175: 21–73.
[36] Sohrabpour, M., A. Kamyab, A. Yari, P. A. Harsini and A. K. Jeihooni. 2024. The factors affecting substance abuse relapse based on theory of planned behavior in male addicts covered by addiction treatment centers in Southern Iran. BMC Public Health 24.
[37] Kan, M. P. and L. R. Fabrigar. 2020. Theory of planned behavior. pp. 5476–5483. In Encyclopedia of Personality and Individual Differences. Cham: Springer International Publishing.

[38] Haviv, N., G. Kaplan Harel, M. Sternfeld and M. Wolff. 2024. Exploring the impact of coercion and voluntary entry on dropout rates in a therapeutic community for substance abuse rehabilitation: A survival analysis and logistic regression study. Probation Journal 02645505231221226.

[39] Bunaciu, A., A. M. Bliuc, D. Best, E. A. Hennessy, M. J. Belanger and C. S. Benwell. 2024. Measuring recovery capital for people recovering from alcohol and drug addiction: a systematic review. Addiction Research and Theory 32(3): 225–236.

[40] Haviv, N., G. Kaplan Harel, M. Sternfeld and M. Wolff. 2024. Exploring the impact of coercion and voluntary entry on dropout rates in a therapeutic community for substance abuse rehabilitation: A survival analysis and logistic regression study. Probation Journal 02645505231221226.

Chapter 10
Brain Chemistry in Common Neurological Disorders

Nicholas B. Wells,[1,*] *Emre Tokgoz,*[2] *Sophia Marra*[1] and *Percin Karakol*[3]

10.1 Disorders

In this chapter we will cover brain-related disorders.

10.1.1 Neurodevelopmental Disorders

Autism Spectrum Disorders are characterized by an inability to adapt to environmental changes and repetitive patterns of behavior and socialization issues. Development of the disorder is not always apparent, and the cause is not entirely clear. Some severe forms of ASD such as childhood disintegrative disorder have been linked to various diseases such as tuberous sclerosis and subacute sclerosing panencephalitis, however the etiology of autism is not well understood. Several of the following studies have attempted to highlight possible neurological foundations for ASD and other neurodevelopmental disorders like ADHD.

Genetic research into the causation of neurodevelopmental disorders is ongoing, but connections are being drawn between specific genetic syndromes and certain psychiatric and neurodevelopmental disorders. Specifically, 22q11.2 deletion

[1] Frank H. Netter School of Medicine, Quinnipiac University, North Haven, CT, 06473, USA.
[2] School of Engineering and Technology, Farmingdale State College, New York, 11735 USA.
[3] Başakşehir Cam and Sakura City Hospital Plastic Reconstructive and Aesthetic Surgery, Istanbul, Turkiye.
* Corresponding author: Nicholas.Wells@quinnipiac.edu

DOI: 10.1201/9781003646044-10

syndrome is associated with increased risk for both autism spectrum disorders, schizophrenia, and attention deficit hyperactivity disorder. 22q11.2 is known as a copy number variant (CNV) syndrome, which draws attention to genetic neural changes in the context of developing a disorder. A study from Doherty et al. [1] highlights abnormal neural and cortical connectivity in connection with 22q11.2 deletion syndrome when compared with a control group. Network function was measured using magnetoencephalography to visualize neural network activity in the two groups. Conclusions suggest a correlation between cortical circuitry due to 22q11.2 deletion as a predisposing factor in the development of neurodevelopmental and psychiatric disorders in children affected by the copy number variant syndrome. Establishment of links between genetic factors and cognitive development remains challenging, but studies such as this suggest that uncovering the mechanisms at work may be on the horizon [2].

While genetic predisposition remains difficult to pin down, the cellular mechanisms of neurodevelopment reveal more concrete underpinnings of brain function in relation to cognitive disorders. Myelin, the sheath formed by oligodendrocytes which conduct action potentials, may play a role in cognitive information processing. Khelfaoui et al. describe the relationship between functional myelin and neurodevelopmental disorders in a review that focuses on myelin's role in memory maintenance and formation. While it is known that myelin serves to conduct nerve impulses, therefore driving prompt thought and action, the correlation with impaired brain function in specific disorders has been revealed but not extensively explored. The specific aspects of myelin that draw interest regarding these disorders are not the fast-conducting properties but the role of myelin in precision movement. Not only is myelin involved with enhancing coordination, but it is also thought that myelin could function to synchronize neural networks and bolster information processing [3].

Looking again at genetics and further into glial cell influence, another recent study on the relationship between neurodevelopmental disorders and genetics by Krzisch et al. describes the effect of glial cell dysregulation on cognitive ability. Neuron-glial interactions have been linked to autism spectrum disorder (ASD) as well as Attention Deficit Hyperactivity Disorder (ADHD) in both human and animal studies. Animal models of ASD have also shown decreased activity of myelin, like the possible connection identified by Khelfaoui et al. between myelin and neurodevelopment. Cell dysregulation, and specifically microglial dysfunction, has been further linked to ASD and ADHD by way of interfering with the excitatory and inhibitory balance within a synapse. Krzisch et al. suggest glial cell dysregulation could be a new target for treatment in addition to neuron-based intervention [4, 5, 6].

10.1.2 Schizophrenia

Schizophrenia is a relatively well-understood condition affecting roughly 1% of the population and involving hallucinations, speech problems, delusions, and unpredictable behavior. There is a myriad of negative symptoms that often accompany the more well-known positive symptoms of schizophrenia such as memory impairment and delayed information processing. Overall, schizophrenia is thought to be caused by a combination of environmental and genetic factors. Most specifically, the dopaminergic and glutaminergic pathways in the brain are heavily involved, like the etiology of addiction. The following section will explore recent literature on the neurological causes of schizophrenia and psychosis.

Schizophrenia treatment focuses mainly on dopaminergic receptor blocking, which has yielded mixed results. D2 receptor blockers have been shown to eventually reach a plateau of efficacy, with 70% receptor saturation being the upper limit for therapeutic effect. Treatment-resistant schizophrenia affects approximately 1/3 of all cases, prompting new research into more precise identification of pathologically defective receptor pathways. A review conducted by Howes et al. delved deep into the neurobiological and neurochemical mechanisms of schizophrenia. Dopaminergic receptors, as the review suggests, may not be entirely at fault for the mechanisms behind schizophrenia. Instead, changes in neurotransmitter prevalence of different types across different brain regions could be driving the clinical presentation of the disease. Specifically, high concentrations of dopamine and glutamine in the basal ganglia are contrasted with reductions in glutamate, GABA, and dopamine levels in the cortex, painting a picture of pronounced imbalance in the schizophrenic brain when compared to an unaffected individual [7, 8].

One of the clinical hallmarks of schizophrenia is poor sleep patterns. From a sleep study perspective, one of the most highly researched areas is sleep spindles. Shown to have effects linked to memory, learning and higher brain function, sleep spindles are suggested as a biomarker of schizophrenia by Ferrarelli et al. Sleep spindles have also been tied to thalamocortical pathway function. The study proposes that not only are sleep spindles both sensitive and specific for schizophrenia, but they may also be a tool for early detection. Sleep spindles may also be indicative of disease progression, as chronic schizophrenic patients showed a more significant reduction in spindle activity compared to new patients. The Ferrarelli et al. study may have laid the foundation for establishing another biomarker for Schizophrenia [9].

Schizophrenia has also been linked to inhibitory neural networks, or a lack of perineuronal nets (PNNs). These perineuronal nets serve to envelope GABAergic interneurons, aiding in development of inhibitory brain pathways. Like some of the other pathophysiological mechanisms of schizophrenia, PNNs involve supporting the function of brain circuitry, 6 of the 7 studies reviewed in an article by Lisboa et al showed a reduction of PNN density in schizophrenic patients, establishing a connection between inhibitory circuit stabilization and schizophrenia. PNNs were also found to be reduced across brain regions, supporting the conclusion that PNNs are the underlying pathophysiological mechanism and not a region-specific consequence of the disease [10].

Despite growing evidence of the neurobiological mechanisms of schizophrenia, the genetic causes remain nebulous. Whole-exome sequencing has produced over 300 genetic variants potentially responsible for the disease. While polygenicity leaves the research community with more questions than answers, twin and family studies have also revealed an environmental component of the disease. Dopamine and glutamate receptors, the targets for most antipsychotics, have been identified as possible genetic markers for the disease. Genome sequencing has revealed many pathological variants, adding to the mystery of schizophrenia while providing a foundation for future genomic analysis [11, 12, 13].

10.1.3 Bipolar Disorder

Bipolar disorder is characterized by chronic, episodic mania and depression. Affected individuals are often treated with a combination of medication and psychotherapy and are highly prone to relapses in symptoms, particularly depressive episodes. The etiology of bipolar disorder is not well understood, but past studies have suggested neurotrophic signaling abnormalities [14, 15].

Bipolar disorder, like many neurological disorders, is thought to occur due to an abnormal inflammatory or immune state [16]. Immune fluctuation states of microglia are thought to even dictate the variations in moods experienced by affected individuals. The adaptive response of microglia to environmental changes such as inflammation and immunity generally helps the brain function, however microglial specialization may also lead to abnormalities seen in neurological immunity that contribute to bipolar disorder. Interactions between genetics and environmental factors contribute to all phases of the disease, and a review by Chaves-Filho et al. outlines some of the more important risk factors in development and progression of Bipolar disorder, as well as the normal processes impacted by the disease. For example, some of the functions of microglia include but are not limited to the following:

- Neurogenesis support
- Phagocytosis
- Immune response and communication
- Signal transmission and astrocyte reactivity
- Neurovascular and blood-brain barrier integrity
- Oligodendrocyte precursor proliferation and differentiation
- Extracellular matrix and synapse remodeling

With the introduction of environmental factors such as childhood trauma, severe or chronic infection, autoimmune disease, and dysbiosis, various states of change occur within microglia. Initially reactive microglia, primed by some environmental and/or genetic factor, can result in manic episodes and subsequent depressive episodes. As the disease progresses, microglia first become hyperreactive, and eventually become dystrophic and degenerate [17, 18].

Grey matter volume was also found to be a crucial factor in the development of bipolar disorder. In a study by Thomas-Odenthal et al., those deemed to be at risk for bipolar disorder, based on genetic and environmental risk factors, as well as affected individuals, were compared with healthy controls. Grey matter volume was found to be higher in certain areas for the affected group, namely the right putamen and left precuneus, while volume was reduced in the right inferior occipital gyrus. While variability in grey matter volume across several regions is informative, the right putamen was deemed as the most significant predictor of bipolar disorder vulnerability [19].

As has been postulated by other literature, Munkholm et al. analyzed 10 studies and uncovered a connection between inflammation and bipolar disorder. Amongst the myriad biomarkers of inflammation, C-reactive protein levels were found to be higher in patients with bipolar disorder when compared with healthy controls [20].

Alongside inflammation, immune cells were also suggested to play a role in bipolar disorder, namely natural killer cells. Neuroinflammation is a hallmark of bipolar disorder as is immune dysregulation, leading to reduced efficacy of both the innate and adaptive immune responses. Specifically, natural killer cells have been identified in bipolar disorder as being permanently activated due to the state of chronic inflammation and autoimmunity found in the disease state. Inhibited immune function of natural killer cells may be contributing to a reduced infection response. The lack of resolution for infectious events thereby is linked to the neuro-immune damage seen over bipolar disorder [21].

10.1.4 Obsessive-compulsive Disorder

Obsessive-compulsive disorder (OCD) is characterized by intensely obsessive behaviors and compulsions that can be debilitating for the affected individual. People suffering from OCD often find themselves in states of distress and disarray if they are unable to complete a compulsive behavior. Obsessive thoughts can intrude and derail normal thought patterns, sometimes not abating until a compulsion is satisfied. OCD affects 1–3% of the global population and has been categorized by the American Psychiatric Association as Obsessive-compulsive and related disorders. OCD has also been related to anxiety, due to several key features of OCD:

- Heightened responsibility
- Overemphasis on thoughts
- Controlling thoughts
- Threat overestimation
- Perfectionism
- Intolerance of uncertainty

The etiology of OCD includes genetic, environmental, and cognitive factors. A mutation in a gene encoding a glutamate transporter, EAAT3, has emerged as a conspirator, according to a 2019 study. Current literature attempts to better understand the altered neurological activity seen in OCD. A recent study involving 95 OCD patients and 106 healthy controls examined differences in neural activity between the two groups. Results showed abnormalities in regional neural connectivity across multiple brain areas when measuring both static and dynamic activity. OCD neural connectivity discrepancies correlated with dopamine, acetylcholine, and norepinephrine transporter activity, suggesting a larger role for neurotransmitters in the disease [22, 23, 24].

A lack of definitive knowledge about the molecular mechanisms driving OCD makes pharmaceutical therapy challenging. Other treatment approaches such as Transcranial direct current stimulation have been utilized with moderate success. A 2024 review detailing treatment for 51 patients resulted in the following outcomes:

- Improved decision making
- Enhanced attention and inhibition
- Reduced attentional bias
- Improved executive control
- Improved working memory

Studies were limited by sample size however the review effectively demonstrates the value of transcranial direct current stimulation as a potential therapy for OCD [25].

While the genetic and molecular causes of OCD are being meticulously investigated, environmental factors affecting development of OCD have been outlined in a recent paper, along with the corresponding brain region impacted by the environmental stressor. Although a direct pathophysiological connection has yet to be made between OCD and stress, early-life stress correlates with changes in goal-directed and habitual behaviors. Stress is known to negatively affect the hippocampus, dorsolateral striatum and amygdala. The dorsal striatum has been found to be hypertrophic in OCD patients, suggesting a relationship between stressful life events and the advancement of OCD. Activity in the dorsal striatum has also been related to compulsive grooming behavior observed in OCD patients. Patients have also reported symptoms beginning during high-stress situations, suggesting not only that stress and OCD share common pathways in the brain, but that they are exacerbated by the same stimuli [26].

From a treatment perspective, while this review will not recommend any specific course, there are psychotherapeutic studies that support regional loci of OCD. Abnormal activation areas were noted across the entire brain, with reduced activation across all lobes following cognitive behavioral therapy and exposure-response prevention [27].

OCD, like the other disorders previously discussed, has been linked to inflammation. Reduced T-regulatory cells and increased T-helper cells in OCD populations points to immune dysregulation as a contributing factor to the disease. OCD has also been associated with increased monocytes that are also more reactive to endotoxin via cytokine release when compared to health controls. The enhanced immune responses driven by increased inflammatory cytokine release and reduced regulation, although not definitive, suggest yet another mental health disorder potentially driven by inflammation. Several interleukins as well as C-reactive protein, a known acute-phase reactant of inflammation, have been implicated in the disease process of OCD. Studies specifically focusing on neuroinflammation have drawn connections between OCD and the release of inflammatory cytokines by activated microglia. Changes in gene expression of monocytes were also observed in OCD patients, indicating a genetic component of disease [28, 29, 30, 31].

10.1.5 Trauma-related Disorders

Post-traumatic stress disorder (PTSD) comes with a host of symptoms that negatively impact mental health. PTSD also carries an increased risk of developing other neurological disorders and even increases susceptibility to suicide. The disorder can develop from a single event or a history of prolonged exposure to negative stimuli. Specific exposures often found to cause PTSD involve violence, injury, sexual trauma, and death. The following review will focus on the neurological effects of PTSD as well as how the brain adapts as a result of the disorder [32].

A major component of PTSD is overactivity of the amygdala. This region of the brain, which is known for the regulation of negative emotion, proves problematic when in a state of dysfunction that leads to a lack of down-regulation of the fear response. The amygdala is thought to be at the mercy of the prefrontal cortex, specifically the ventral medial region, and responds to threatening stimuli. A lack of extinction when it pertains to a fear or anxiety response is characteristic of the amygdala in PTSD. Promising results have been seen in labeling therapy, which involves verbally categorizing a stimulus with the emotional response that it provokes. By labeling a threatening stimulus, healthy individuals have shown the ability to down-regulate the fear response driven by the amygdala. Combat-related PTSD patients also showed reductions in reactivity to traumatic combat images following affect labeling intervention, further solidifying the downregulation of the amygdala response as a possible solution to the debilitating disorder [33, 34].

There is discussion as well about other treatment approaches that take advantage of latent-state and model-based learning. Damage to the orbitofrontal cortex could explain the latent-state nature of PTSD, which drives many of the issues seen in the disorder regarding extinction of fear. With latent learning, the subconscious gathers information about a stimulus or event and later uses that previously obtained information to garner a response. In PTSD, this style of learning inhibits the affected individual's ability to eliminate the unwanted response. The negative response to a certain is expected and therefore is more likely to occur compared to behavioral extinction. Difficulty with learning and appraisal in PTSD patients can partially be explained by changes in neurocircuitry.

As outlined in Figure 1, the ACC and amygdala pathways are upregulated in PTSD, while the striatal, hippocampal and prefrontal pathways are downregulated. Threat detection is compromised via heightened interactions between the dorsal anterior cingulate cortex, amygdala, and anterior insula. This creates a state of arousal prone to fear and anxiety. A reduction in updating task representation also leads to rigidity in thought, allowing for less natural fear extinction compared to healthy controls. Most thought regarding representation occurs in the hippocampus and prefrontal cortex. The dorsal and ventral striatum normally function to process reward and therefore guide behavior; however, this pathway is downregulated in PTSD, resulting in less reward seeking/learning.

Frontoparietal activity is thought to underlie task representation along with the hippocampus, explaining the memory and attention deficits experienced by PTSD patients. In exposure to chronic stress, there is a lack of proper discrimination by the hippocampus between new and previously encountered stimuli [35, 36, 37].

Impact of trauma and stressor exposure on learning and appraisal pathways in PTSD		
Neurocircuitry	**Higher-order function**	**Lower-order function**
Dorsal and Ventral Striatum	Reward processing and action	Reward learning, reward seeking, positive affect
Lateral PFC, hippocampus, ventromedial PFC	Updating task representation	Fear extinction, re-appraisal, context processing
ACC and amygdala	Threat detection	Hypervigilance, hyperarousal, fear learning, anxiety

Fig. 1. A summary of trauma and stressor factors' impact on learning.

Stress and working memory share a close association. Chronic stress, such as that seen in PTSD, has been linked to a reduction in working memory due to the inability to discard negative thoughts or sift aside information that is not relevant to the task at hand. Impaired discrimination between stimuli within the realm of working memory also explains why PTSD patients show a tendency toward improper threat appraisal [38].

A handful of genetic factors have been found that could be associated with adversity management. Low resilience is associated with PTSD and has also been linked to alterations in the expression of immune and dopamine genes. On the contrary, individuals with higher resilience to traumatic events were shown to have a less significant inflammatory response. Two specific genes involved encode a serotonin transporter and neuropeptide Y, which is a critical protein in hypothalamic-pituitary-adrenal axis function. While a gene-environment interaction or a precise genetic marker have not been identified, the neurocircuitry of PTSD coupled with the emergence of several promising biomarkers bodes well for the future understanding and treatment of the disorder [39, 40, 41].

10.1.6 Dissociative Identity Disorder

Dissociative identity disorder (DID) affects 1.5% of the global population and is characterized by severe behavioral health symptoms, including a predisposition toward self-harm and suicide. Associated with demonic possession in some corners of the world, DID is typically linked to extensive abuse or trauma experienced during childhood. The disease involves switching between multiple different personalities that have little to no association with one another which develop as a coping mechanism to alleviate an absent sense of self. Identity switching is generally preceded by a seizure-like episode and is usually accompanied by amnesia and memory loss [42, 43].

The seizure-like state during identity exchange is a hallmark of the clinical presentation. A recent review revealed that several structures within the brain have been implicated in DID, specifically the amygdala, the orbitofrontal cortex, and the anterior cingulate cortex. Brain imaging studies showed increased resting state activity between the amygdala and middle/inferior frontal gyrus, contrasted with reduced reactivity to stress in the amygdala and hippocampus. In combination, the

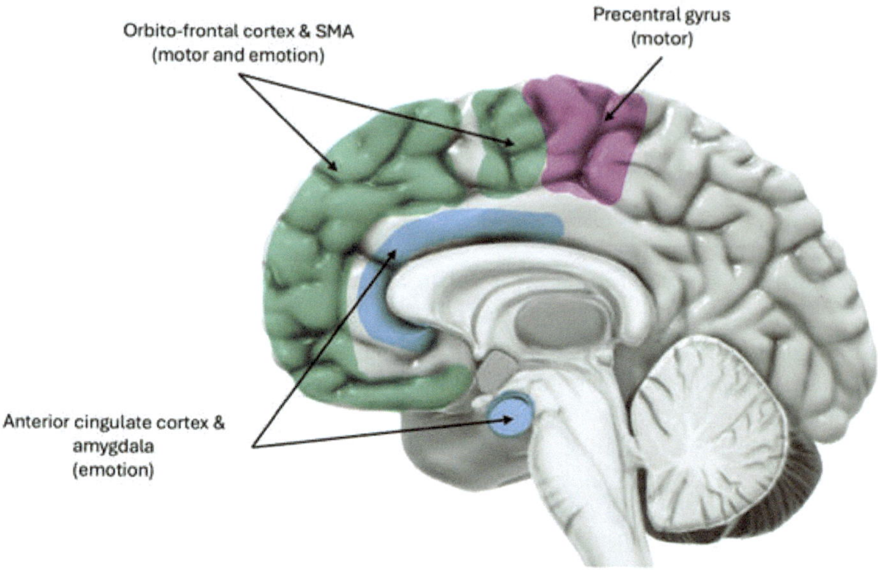

Fig. 2. Regions of the brain with motor and emotion control locations affected.

hyper- and hyporesponsive nature of these brain regions highlights the dysregulation of a proper stress response in DID patients compared to healthy controls. Figure 2 is a demonstration of the corresponding regions observed to be affected in the brain.

The same review proposed a reduced connectivity between regions involved in motor control and cognition while also indicating heightened connectivity regarding emotional processing. The orbitofrontal cortex, amygdala, and anterior cingulate cortex remain the most frequently referenced regions when discussing dysfunction in the DID brain [44, 45].

Other studies have also postulated dysfunctional limbic system function alongside structural changes in the parietal and occipital lobes of the brain. One study also suggests glutamate transport could be a significant contributor to the symptoms associated with dissociative episodes that plague DID patients. Overall, brain imaging has proved effective in highlighting regions with hyper- and hypo-reactivity in DID patients. More specific regional and neurobiological findings will continue to aid in the understanding and treatment of this condition [46, 47, 48, 49].

10.2 Depression

10.2.1 Clinical Features

One of the most common mental health disorders with a prevalence of 7% worldwide, depression is characterized by general feelings of emptiness, sadness and irritability alongside physical and mental changes that combine into a formidable disorder. 60% of those affected by depression go untreated, due in part to stigma as well as skepticism surrounding medication. A familial disposition is suggested, as people with immediate family members who suffer from depression are nearly three-fold more likely to be diagnosed than the average person. The scope of depression is large, encompassing 5 different but related disorders [50]:

1. Disruptive mood dysregulation disorder
2. Major depressive disorder
3. Persistent depressive disorder
4. Premenstrual dysphoric disorder
5. Depressive disorder due to another medical condition

The family of depressive disorders deserves a robust neurobiological review; however, this review will focus on brain chemistry and neurobiology in the broader category of depression.

10.2.2 Brain Chemistry in Depression

It has long been proposed that depression is caused by a serotonin imbalance in the brain. Initial treatment approaches targeted serotonin, but the effects of the medication also impacted levels of dopamine and noradrenaline. Research evolved to include all monoamines in the hypothesis of neurotransmitter deficiency being the heart of depression. The modern era has seen the emergence of serotonin reuptake inhibitors and serotonin noradrenaline reuptake inhibitors, two medications designed to increase synaptic interaction of critical neurotransmitters. There has even recently been an introduction of classically recreation drugs into the depression treatment space. Ketamine, LSD, and DMT have all been used in clinical trials to treat depression, extending the scope of treatment beyond serotonin and into the dopamine and glutamine realm.

Four main neural networks have been associated with depression. The affective network, reward network, cognitive control network, and default mode network. The areas most seen with abnormal connectivity are the amygdala, hippocampus, and prefrontal cortex. The frontal regions of the brain are closely intertwined with the serotonin pathway while the reward network is built on the foundation of the dopaminergic pathway.

The whole picture of depression shows several areas of the brain displaying a heterogenous degree and direction of affect. With multiple interconnected axes in disarray, it is easy to see how complex depression treatment has become. Future studies are prioritizing a deeper understanding of the serotonin pathway and a more robust understanding of 5-HT release and receptor binding [51, 52].

One article refers specifically to the metabolism of neurotransmitters when discussing depression. The metabolic pathways of patients with major depressive disorders were found to be lower for GABA, dopamine, and tryptophan compared to healthy controls. Normal function of the monoaminergic system allows for mood stability and motivation. By-products of monoamine metabolism have been found to be higher in individuals with major depressive disorder, suggesting a higher neurotransmitter turnover rate that could be driving the negative symptoms of low energy and motivation associated with depression [53].

Hypotheses about depression causation in the scientific community include dopamine, serotonin, glutamate, and even histamine as factors. Electrochemical techniques have been used to elicit a clearer picture of the impact of pharmaceuticals on these neurotransmitter pathways. By measuring real-time activity and medication-induced changes, researchers can observe the mechanism of action of a drug on its neurochemical target. Histamine as a mediator of inflammation has become another target for therapy. When given in conjunction with serotonin reuptake inhibitors, histamine metabolism inhibitors aided in increasing serotonin concentrations in mice. Histamine receptors were also found to release and regulate other neurotransmitters in the central nervous system, namely serotonin, dopamine, acetylcholine, GABA, and norepinephrine. The powerful H3 receptor provides yet another link between inflammatory response and mental health disorders [54].

10.3 Generalized Anxiety Disorder

10.3.1 Clinical Features

Anxiety, classified more specifically as Generalized Anxiety Disorder, is a common mental health disorder affecting approximately 20% of adults worldwide. The defining characteristic of the disease is a near-constant state of worry. In most cases, the state of worry or panic is excessive and unfounded. Some key symptoms of the disease are:

- Restlessness or irritability
- Trouble sleeping
- Being easily fatigued
- Muscle tension

Compared to healthy people, a hallmark of anxiety is a worry or fear that is not based on any tangible stimulus. Anxiety creates a state of distress, especially in new or challenging situations. Anxiety has a relatively unknown etiology, although studies have postulated a link between disorder and serotonin and/or norepinephrine. Thus, selective serotonin and norepinephrine reuptake inhibitors have emerged as the go-to pharmaceutical therapy [55].

10.3.2 Brain Chemistry in Anxiety

Imaging of the brain has been attempted to ascertain the regions responsible for the pathological behaviors of anxiety. Though no sole region has been identified as the culprit, a connection between several regions has been identified. Anxiety may be related to a faulty circuit that spans many brain regions, specifically the amygdala, hypothalamus, hippocampus, medial prefrontal cortex, and anterior cingulate cortex. Since generalized anxiety disorder involves a constant state of worry without cause, the brain areas responsible for identifying actual causes of distress ought naturally to be abnormal in the affected brain. Fear and anxiety responses can differ based on the stimulus, further complicating the already daunting task of pinning anxiety to a specific brain locus [56].

It has been suggested that patients suffering from GAD have alternations in grey matter in various regions. Multiple studies reference different brain regions with reduced grey matter. In an attempt to home in on a consensus, Ou et al. analyzed several voxel-based morphometry (VBM) studies. VBM teases out differences in grey matter, white matter, and cerebrospinal fluid within the brain. The result was reduced grey matter in the left superior temporal gyrus as well as significant association between anxiety severity and altered grey matter within the right striatum and reticular gyrus. Anxiety severity was also correlated with grey matter alterations in the insula, putamen, and striatum. In children, the superior temporal gyrus showed asymmetry in patients with GAD. The superior temporal gyrus plays an important role in representation of emotions, adding to the significance of the findings. It remains unclear whether a single region is at fault for GAD, however, many studies have provided convincing evidence of some regions likely to be at least partially responsible [57].

Biomarkers have been investigated to better understand the etiology of anxiety. There are several proteins that are upregulated during periods of intense stress. When observing neural distress, the S100B protein concentration is a reliable benchmark. The protein exists in astrocytes and glial cells and plays many roles within the central nervous system including cell growth, enzymatic activity, and electrolyte homeostasis. S100B is shown to be elevated in times of stress and is also neurotoxic at high concentrations despite having positive neurotrophic effects at low concentrations. S100B was found in elevated concentrations in patients with GAD, as was 8-Hydroxy-2'-deoxyguanosine (8-OHdG), which is a marker for neural oxidative stress due to reactive oxygen species. The survival and proliferative characteristics of S100B provide insight into how the stress response is so tightly intertwined with anxiety while the 8-OHdG elevations confirm the cytotoxic effects of stress due to reactive oxygen species [58, 59].

A study aimed at identifying biomarkers of anxiety identified 17 epigenetic markers. The two major categories of biomarkers implied in anxiety disorders were involved with mitochondrial dysfunction, apoptosis, and neurosignaling. (Kwon) Another study looked into interleukin levels via enzyme-linked immunosorbent assay (ELISA) in GAD patients compared to healthy controls and found significantly elevated levels of IL-17A and IL-23A, two key mediators of pro-inflammation. IL-23A drives activation of Th17 helper cells while IL-17A is an inflammatory cytokine. Stress-induced anxiety was correlated with increased levels of cytokines overall, and links have been established between severity of GAD and certain pro-inflammatory biomarkers. Thus, inflammation returns as a key suspect as we understand more about GAD and other mental health disorders [60, 61].

References

[1] Doherty, J. L., A. C. Cunningham, S. J. Chawner, H. M. Moss, D. C. Dima, D. E. Linden et al. 2024. Atypical cortical networks in children at high-genetic risk of psychiatric and neurodevelopmental disorders. Neuropsychopharmacology 49(2): 368–376.

[2] Mughal, S., R. M. Faizy and A. Saadabadi. 2024. Autism Spectrum Disorder. [Updated 2022 Jul 19]. In: StatPearls [Internet]. Treasure Island (FL): StatPearls Publishing. Available from: https://www.ncbi.nlm.nih.gov/books/NBK525976/.

[3] Khelfaoui, H., C. Ibaceta-Gonzalez and M. C. Angulo. 2024. Functional myelin in cognition and neurodevelopmental disorders. Cellular and Molecular Life Sciences 81(1): 1–14.

[4] Krzisch, M., S. D'Antoni and L. Ciranna. 2024. Role of glia in neurodevelopmental disorders. Frontiers in Cellular Neuroscience 18: 1404362.

[5] Bennett, M. L. and A. N. Viaene. 2021. What are activated and reactive glia and what is their role in neurodegeneration? Neurobiology of Disease 148: 105172.

[6] Santos da Silva Calado, C. M., R. Manhães-de-Castro, V. D. S. Souza, H. J. Cavalcanti Bezerra Gouveia, S. D. C. Pereira, M. M. da Silva et al. 2024. Early-life malnutrition role in memory, emotional behavior and motor impairments in early brain lesions with potential for neurodevelopmental disorders: a systematic review with meta-analysis. Nutritional Neuroscience 1–23.

[7] Hany, M., B. Rehman, A. Rizvi and J. Chapman. 2025. Schizophrenia. [Updated 2024 Feb 23]. StatPearls [Internet]. Treasure Island (FL): StatPearls Publishing.

[8] Howes, O. D., B. R. Bukala and K. Beck. 2024. Schizophrenia: from neurochemistry to circuits, symptoms and treatments. Nat Rev Neurol. 20: 22–35. https://doi.org/10.1038/s41582-023-00904-0.

[9] Ferrarelli, F. 2024. Sleep spindles as neurophysiological biomarkers of schizophrenia. European Journal of Neuroscience 59(8): 1907–1917.

[10] Lisboa, J. R. F., O. Costa, G. H. Pakes, D. A. E. Colodete and F. V. Gomes. 2024. Perineuronal net density in schizophrenia: A systematic review of postmortem brain studies. Schizophrenia Research 271: 100–109.

[11] Sullivan, P. F., S. Yao and J. Hjerling-Leffler. 2024. Schizophrenia genomics: genetic complexity and functional insights. Nature Reviews Neuroscience 1–14.

[12] Krcmar, M., C. M. Wannan, S. Lavoie, K. Allott, C. G. Davey, H. P. Yuen et al. 2024. The self, neuroscience and psychosis study: Testing a neurophenomenological model of the onset of psychosis. Early Intervention in Psychiatry 18(2): 153–164.

[13] Tonna, M. 2024. The evolution of symbolic thought: at the intersection of schizophrenia psychopathology, ethnoarchaeology, and neuroscience. Culture, Medicine, and Psychiatry 1–18.

[14] Jain, A. and P. Mitra. 2024. Bipolar Disorder. [Updated 2023 Feb 20]. In: StatPearls [Internet]. Treasure Island (FL): StatPearls Publishing; Available from: https://www.ncbi.nlm.nih.gov/books/NBK558998/.

[15] Scaini, G., S. S. Valvassori, A. P. Diaz, C. N. Lima, D. Benevenuto, G. R. Fries et al. 2020. Neurobiology of bipolar disorders: a review of genetic components, signaling pathways, biochemical changes, and neuroimaging findings. Braz J. Psychiatry 42(5): 536–551.

[16] Konopaske, G. T., N. Lange, J. T. Coyle and F. M. Benes. 2014. Prefrontal cortical dendritic spine pathology in schizophrenia and bipolar disorder. JAMA Psychiatry 71(12): 1323–31.

[17] Chaves-Filho, A., C. Eyres, L. Blöbaum, A. Landwehr and M. È. Tremblay. 2024. The emerging neuroimmune hypothesis of bipolar disorder: an updated overview of neuroimmune and microglial findings. Journal of Neurochemistry.

[18] Toomukuntla, S., C. V. Vemula, M. S. Spoorthy, S. A. Zaki and S. K. Tikka. 2024. Prevalence and risk of schizophrenia and bipolar disorder in patients with type 1 diabetes mellitus: a systematic review and meta-analysis. Indian Journal of Psychological Medicine 02537176241238959.

[19] Thomas-Odenthal, F., F. Stein, C. Vogelbacher, N. Alexander, A. Bechdolf, F. Bermpohl et al. 2024. Larger putamen in individuals at risk and with manifest bipolar disorder. Psychological Medicine 1–11.

[20] Munkholm, K., I. J. O. Mäkinen, K. Maigaard, K. Coello, A. K. Pagsberg and L. V. Kessing. 2024. Inflammatory and oxidative stress biomarkers in children and adolescents with bipolar disorder–a systematic review and meta-analysis. Neuroscience and Biobehavioral Reviews 105766.

[21] Tamouza, R., M. Leboyer and V. Vieillard. 2024. Natural killer (NK) cells in bipolar disorders. Neuroscience Applied 104066.

[22] Brock, H., A. Rizvi and M. Hany. 2024. Obsessive-Compulsive Disorder. [Updated 2024 Feb 24]. In: StatPearls [Internet]. Treasure Island (FL): StatPearls Publishing. Available from: https://www.ncbi.nlm.nih.gov/books/NBK553162/.

[23] Escobar, A. P., J. R. Wendland, A. E. Chávez and P. R. Moya. 2019. The neuronal glutamate transporter EAAT3 in obsessive-compulsive disorder. Front Pharmacol. 10: 1362.

[24] Tao, Q., J. Dang, H. Guo, M. Zhang, X. Niu, Y. Kang et al. 2024. Abnormalities in static and dynamic intrinsic neural activity and neurotransmitters in first-episode OCD. Journal of Affective Disorders.

[25] Spiroiu, F., L. Minuzzi, D. Duarte, R. McCabe and N. Soreni. 2024. Neurocognitive effects of transcranial direct current stimulation in obsessive-compulsive disorder: a systematic review. International Journal of Neuroscience, (just-accepted) 1–17.

[26] Debnath, A. and S. Sarkar. 2024. Impact of early life stress on neurobiological vulnerability to obsessive-compulsive disorder: early life stress and neurobiological OCD risk. Brain Science Advances 10(2): BSA-2024.

[27] Stephenson, C., A. Philipp-Muller, E. Moghimi, J. Y. Nashed, D. J. Cook, A. Shirazi et al. 2024. Effects of cognitive behavioural therapy and exposure–response prevention on brain activation in obsessive–compulsive disorder patients: systematic review and meta-analysis. European Archives of Psychiatry and Clinical Neuroscience 1–17.

[28] Coelho, D. R. A., J. D. Salvi, W. F. Vieira and P. Cassano. 2024. Inflammation in obsessive–compulsive disorder: A literature review and hypothesis-based potential of transcranial photobiomodulation. Journal of Neuroscience Research 102(3): e25317.

[29] Pankratz, B., K. von Zedtwitz, K. Runge, D. Denzel, K. Nickel, A. Schlump et al. 2023. Cerebrospinal fluid findings in adult patients with obsessive-compulsive disorder: A retrospective analysis of 54 samples. The World Journal of Biological Psychiatry 24(4): 292–302.

[30] Li, J., J. Cheng, L. Yang, Q. Niu, Y. Zhang and L. Palaniyappan. 2024. Association of cortical gyrification, white matter microstructure, and phenotypic profile in medication-naïve obsessive–compulsive disorder. Psychological Medicine 54(8): 1573–1579.

[31] Becker, H. C., A. M. Beltz, J. A. Himle, J. L. Abelson, S. R. Block, S. F. Taylor et al. 2024. Changes in brain network connections after exposure and response prevention therapy for obsessive-compulsive disorder in adolescents and adults. Biological Psychiatry: Cognitive Neuroscience and Neuroimaging 9(1): 70–79.

[32] Mann, S. K., R. Marwaha and T. J. Torrico. 2024. Posttraumatic Stress Disorder. [Updated 2024 Feb 25]. *In*: StatPearls [Internet]. Treasure Island (FL): StatPearls Publishing. Available from: https://www.ncbi.nlm.nih.gov/books/NBK559129/.

[33] Burklund, L. J., C. D. Davies, A. Niles, J. B. Torre, L. Brown, M. Vinograd et al. 2024. Affect labeling: a promising new neuroscience-based approach to treating combat-related PTSD in veterans. Frontiers in Psychology 15: 1270424.

[34] Aupperle, R. L., A. J. Melrose, M. B. Stein and M. P. Paulus. 2012. Executive function and PTSD: Disengaging from trauma. Neuropharmacology 62(2): 686–694.

[35] Wu, C. T., D. Haggerty, C. Kemere and D. Ji. 2017. Hippocampal awake replay in fear memory retrieval. Nature Neuroscience 20(4): 571–580.

[36] Resick, P. A., T. E. Galovski, M. O. B. Uhlmansiek, C. D. Scher, G. A. Clum and Y. Young-Xu. 2008. A randomized clinical trial to dismantle components of cognitive processing therapy for posttraumatic stress disorder in female victims of interpersonal violence. Journal of Consulting and Clinical Psychology 76(2): 243.

[37] Cisler, J. M., J. E. Dunsmoor, G. A. Fonzo and C. B. Nemeroff. 2024. Latent-state and model-based learning in PTSD. Trends in Neurosciences.

[38] Almarzouki, A. F. 2024. Stress, working memory, and academic performance: a neuroscience perspective. Stress 27(1): 2364333.

[39] Gan, Y., H. Huang, X. Wu and M. Meng. 2024. What doesn't kill us makes us stronger: insights from neuroscience studies and molecular genetics. Current Opinion in Behavioral Sciences 59: 101431.

[40] Mehta, D., O. Miller, D. Bruenig, G. David and J. Shakespeare-Finch. 2020. A systematic review of DNA methylation and gene expression studies in posttraumatic stress disorder, posttraumatic growth, and resilience. Journal of Traumatic Stress 33(2): 171–180.
[41] Yunus, A. 2023. Adverse childhood experiences: the neuroscience of trauma, resilience, and healing throughout the life course: by Kathleen Brewer-Smyth, 2022, Cham, Springer, xxv + 454 pp., ISBN: 978-3-031-08801-8, (eBook) €50.28. Education 3-13, 52(2): 288–290. https://doi.org/10.1080/03004279.2023.2183082.
[42] Mitra, P. and A. Jain. 2023. Dissociative Identity Disorder. [Updated 2023 May 16]. In: StatPearls [Internet]. Treasure Island (FL): StatPearls Publishing; Available from: https://www.ncbi.nlm.nih.gov/books/NBK568768/.
[43] Maldonado, J. R. and D. Spiegel. 2023. Dissociative Disorders. *In*: Tasman, A. et al. Tasman's Psychiatry. Springer, Cham. https://doi.org/10.1007/978-3-030-42825-9_80-1.
[44] Hassan, J., S. Taib and A. Yrondi. 2024. Structural and functional changes associated with functional/dissociative seizures: A review of the literature. Epilepsy and Behavior 152: 109654.
[45] Mark, J., Q. Shoukat, J. Bayer and E. Harris. 2024. Pathophysiology of dissociative identity disorder. pp. 141–144. In Dissociative Identity Disorder: Treatment and Management. Cham: Springer International Publishing.
[46] Hasan, F. M., N. Bonsu and V. Sreeram. 2024. Dissociative identity disorder and the human brain; neuroanatomy. pp. 137–140. In Dissociative Identity Disorder: Treatment and Management. Cham: Springer International Publishing.
[47] Purcell, J. B., B. Brand, H. A. Browne, R. A. Chefetz, M. Shanahan, Z. A. Bair et al. 2024. Treatment of dissociative identity disorder: leveraging neurobiology to optimize success. Expert Review of Neurotherapeutics 24(3): 273–289.
[48] Kar, S. 2024. Different kinds of dissociative disorders, including dissociative identity disorder. pp. 21–32. In Dissociative Identity Disorder: Treatment and Management. Cham: Springer International Publishing.
[49] Jamieson, G., E. Cardeña and V. de Pascalis. 2024. A spontaneous dissociative episode during an EEG experiment. Brain and Cognition 174: 106121.
[50] Chand, S. P. and H. Arif. Depression. [Updated 2023 Jul 17]. In: StatPearls [Internet]. Treasure Island (FL): StatPearls Publishing; 2024 Jan-. Available from: https://www.ncbi.nlm.nih.gov/books/NBK430847/.
[51] Bremshey, S., J. Groß, K. Renken and O. A. Masseck. 2024. The role of serotonin in depression—A historical roundup and future directions. Journal of Neurochemistry.
[52] Moncrieff, J., R. E. Cooper, T. Stockmann, S. Amendola, M. P. Hengartner and M. A. Horowitz. 2024. Difficult lives explain depression better than broken brains. Molecular Psychiatry 1–4.
[53] Zhao, T., T. Liu, L. Wang, K. Xie, H. Tang and M. Tang. 2024. Dysfunction of neurotransmitter metabolism is associated with the severity of depression in first-diagnosed, drug-naïve depressed patients. Journal of Affective Disorders 349: 332–341.
[54] Dunham, K. E. and B. J. Venton. 2024. Electrochemical and biosensor techniques to monitor neurotransmitter changes with depression. Analytical and Bioanalytical Chemistry 416(9): 2301–2318.
[55] Munir, S. and V. Takov. 2022. Generalized Anxiety Disorder. [Updated 2022 Oct 17]. In: StatPearls [Internet]. Treasure Island (FL): StatPearls Publishing; 2024 Jan-. Available from: https://www.ncbi.nlm.nih.gov/books/NBK441870/.
[56] Drzewiecki, C. M. and A. S. Fox. 2024. Understanding the heterogeneity of anxiety using a translational neuroscience approach. Cognitive, Affective, and Behavioral Neuroscience 24(2): 228–245.
[57] Ou, C. H., C. S. Cheng, P. L. Lin and C. L. Lee. 2024. Grey matter alterations in generalized anxiety disorder: A voxel-wise meta-analysis of voxel-based morphometry studies. International Journal of Developmental Neuroscience 84(4): 281–292.
[58] Oktay, M., M. Asoğlu, S. Taskin and A. Kirmit. 2024. Biological markers in newly diagnosed generalized anxiety disorder patients: 8-OHdG, S100B and oxidative stress. Neuropsychiatric Disease and Treatment 19–24.
[59] Zhang, Z. 2024. A comprehensive overview on the generalized anxiety disorder–etiology and treatment. In SHS Web of Conferences (Vol. 193, p. 03008). EDP Sciences.

[60] Kwon, Y., A. Blazyte, Y. Jeon, Y. J. Kim, K. An, S. Jeon et al. 2024. Identification of 17 novel epigenetic biomarkers associated with anxiety disorders using differential methylation analysis followed by machine learning-based validation. bioRxiv, 2024-05.

[61] Roknuzzaman, A. S. M., R. Sarker, J. Nayem, M. A. Bhuiyan, M. R. Islam and Z. Al Mahmud. 2024. Altered serum interleukin 17A and interleukin-23A levels may be associated with the pathophysiology and development of generalized anxiety disorder. Scientific Reports 14(1): 15097. Chicago.

Chapter 11

The Brain on Controlling Attention and Managing Anxiety

Percin Karakol,[1,*] *Ceren Comert,*[1] *Sophia Marra*[2] and *Emre Tokgoz*[3]

11.1 Introduction

Anxiety disorders encompass a range of conditions that exhibit similar symptoms and develop within a specific context [1, 2]. The relationship between normal anxiety and pathological anxiety can be understood through disruptions in neurocognitive elements such as attention. Research on attention particularly presents a valuable opportunity to comprehend cognitive functioning in individuals with anxiety disorders. The term "attention" encompasses a group of cognitive processes that enable the brain to focus on specific stimuli for more detailed analysis.

Much of theoretical in this area of research interest defines anxiety as a motivational state activated by a heightened perception of threat to the individual [3]. In these situations, it has been observed that state anxiety that relies on a temporary and situational feeling of nervousness, worry, or apprehension can hinder cognitive performance, affecting processes such as working memory and the distribution of attentional resources. Studies on attentional bias suggest that people with either high state or trait anxiety tend to direct their attention more towards threatening stimuli rather than neutral ones, and this tendency is stronger than that seen in individuals who do not experience anxiety or in low-anxiety scenarios [4–7]. Nevertheless, the

[1] Başakşehir Cam and Sakura City Hospital Plastic Reconstructive and Aesthetic Surgery, Istanbul, Turkiye.

[2] Frank H. Netter School of Medicine, Quinnipiac University, North Haven, CT, 06473, USA.

[3] School of Engineering and Technology, Farmingdale State College, New York, 11735 USA.

* Corresponding author: ppercin@gmail.com

DOI: 10.1201/9781003646044-11

influence of attentional resource allocation may change dynamically throughout different stages of neural processing [8–11].

Attention is proposed to consist of multiple processes in [33]. The initial response to threatening information—vigilance—is thought to be governed by a "bottom-up" system. Once a threat captures attention, individuals may either swiftly disengage from it or remain focused on it that can lead to difficulty in shifting attention away. This engagement or avoidance is believed to be managed by "top-down" processes. According to these perspectives, an individual's ability to regulate attention through top-down mechanisms likely affects the extent of their attentional biases, especially in contexts where realistic threats might render these adaptive biases [12]. Considering the intricacies of how anxiety affects attention to threats, attentional bias might be more accurately viewed as the result of multiple processes operating within an attentional framework rather than as a single phenomenon.

Although a consistent finding is the presence of an anxiety-related threat bias, both classic and contemporary studies have highlighted broader deficits in attentional and cognitive control linked to this personality trait, even in tasks that do not involve emotional content. Such deficits have often been noted in areas such as language comprehension, as discussed in other reviews [13].

In relation to attention, research indicates that anxiety is connected to prolonged reaction times in antisaccade tasks that require participants to suppress a sudden peripheral target while intentionally shifting their focus to the opposite side [3–14]. The influence of anxiety is also evident in the heightened processing of irrelevant distractors, including competing words, letters, and arrows [15–17]. Furthermore, there is a noted decrease in negative priming referring to the inhibition of target locations that were previously occupied by distractors [16].

In this work, we summarize recent research regarding the connection between attention and anxiety. As previously mentioned, this paper aims to build upon earlier reviews by providing a more focused and in-depth exploration of attentional orientation within anxiety disorders.

11.2 Anxiety

Anxiety is an emotion associated with negative effects that negatively impacts performance from a cognitive psychological standpoint [18, 19]. It is expressed as a state of distress, tension, and heightened vigilance in response to perceived threats. While anxiety can serve adaptive functions by preparing individuals for action, increasing arousal, and enhancing bottom-up sensory processing, it can also hinder top-down cognitive control. This impairment often arises from distractions caused by both internal and external stimuli that are irrelevant to the task at hand that can detrimentally affect cognitive functioning [4–13].

Many theorists propose that detrimental performance effects are linked to anxiety primarily stemming from the way worry and other forms of cognitive interference, such as self-focus, consume attention [22]. These distraction or attentional theories posit that a crucial component of effective task performance is the ability to focus on task-relevant cues, processes, and behaviors while disregarding information that is either irrelevant or disruptive [23]. When these attentional processes are interfered with, it is assumed that performance suffers, and in severe instances, this can lead to what is known as "choking" referring to a significant decline in performance when the stakes for success are high [24].

In addition to widely recognized anxiety-related threat biases, both classic and modern studies have highlighted broader attentional and cognitive control deficits connected to this personality trait, even in non-emotional tasks. These deficits have commonly been observed in areas such as language comprehension, as reviewed in other literature [13]. Anxiety is identified to be associated with longer reaction times during antisaccade tasks requiring suppressing a sudden peripheral target and intentionally shifting focus to the opposite side of that target [3–14]. Influence of anxiety is further reflected in the heightened processing of irrelevant distractions like competing words, letters, or arrows [15–17] as well as decreased negative priming helping inhibit target locations that were previously filled by distractors [16].

Furthermore, brain imaging studies can reveal the neural mechanisms that underlie orienting behaviors. Activation in certain brain regions corresponds with processes that support attentional biases, demonstrating a positive correlation with anxiety symptom levels. Conversely, activation in other regions reflects compensatory processes, leading to negative correlations with anxiety symptoms. Therefore, attention orienting behavior represents the final behavioral output resulting from a complex and well-coordinated interaction among various neural components in a circuit.

11.3 Attention Control Theory

The attentional system can be categorized into two main subsystems: a top-down subsystem that is goal-directed and volitional, and a bottom-up subsystem that is stimulus-driven and reflexive. According to ACT, anxiety disrupts the equilibrium between these two subsystems by diminishing the role of top-down processes and, as a result, amplifying the influence of bottom-up, stimulus-driven processes. Numerous studies using different methodologies provide substantial evidence that anxiety adversely impacts the efficiency of crucial central executive functions within working memory, particularly in areas such as inhibition, shifting, and updating information. This impairment subsequently hinders goal-directed behavior and reduces attentional control. Building on the primary tenets of ACT, it is essential to identify a causal mechanism through which decreased attentional control can intensify the negative effects of anxiety on performance outcomes, particularly by fostering a focus on ruminative and worrisome thoughts. One notable consequence of diminished attentional control is the 'hidden' cost associated with compensatory strategies that high-anxiety individuals employ to maintain performance, which ultimately amplifies the impact of anxiety on processing efficiency. Systematic behavior of anxiety-prone individuals occurs that often find their worries compete for their mental resources. This competition undermines their ability to control attention, leading to higher levels of distractibility and a reinforcement of their worrying thoughts. In other words, there is a reciprocal relationship where worry affects attention control, and attentional control influences worry [25].

ACT suggests that increasing levels of anxiety, whether state or trait, result in diminished attentional control. This shift means that the influence of the goal-directed system is weakened, while the stimulus-driven system gains strength, particularly through the preferential processing of task-irrelevant stimuli related to threats. ACT posits that insufficient attention control could act as a risk factor for developing high trait anxiety, a notion that is further clarified by the moderation model. Although the directionality of this relationship is still a topic of debate, it is reasonable to consider that there are bidirectional effects, as state anxiety during evaluation can notably disrupt immediate task performance in terms of both effectiveness and efficiency.

Additionally, ACT acknowledges that attentional control is a complex construction, and the effects of anxiety may vary across different components and indicators of attentional control. Under certain conditions, individuals with high trait anxiety might not demonstrate overt disruptions in attentional control based on response accuracy. However, achieving this level of performance could require a greater allocation of cognitive resources compared to their low trait anxious counterparts that may depend on their motivation to excel. Consequently, ACT predicts that the attentional impairments associated with anxiety may lead to decreased efficiency, characterized by slower response rates on cognitive tasks when motivated, without significantly affecting the overall effectiveness or accuracy of their responses.

Despite some support for this prediction, few studies have directly compared task effectiveness and efficiency. Moreover, the effects of diminished cognitive efficiency might only be evident under high cognitive load conditions, where opportunities to engage additional cognitive resources are restricted. Thus, researchers can assess the relationship between task effectiveness, efficiency, and attentional control by manipulating cognitive load in experimental designs.

In summary, ACT posits that both high trait and state anxiety can adversely impact the efficiency of processing related to shifting and inhibition functions. It is also suggested that anxiety may only affect the updating function when cognitive load is high. Nonetheless, the evidence supporting these hypotheses has been inconsistent.

In recent research exploring the cognitive effects of anxiety, a series of experiments are conducted in [55] that investigated the differences between high and low trait anxious individuals under conditions of high and low threat, or state anxiety. Their findings indicated that trait anxiety was linked to reduced efficiency and effectiveness in inhibition tasks, while state anxiety did not appear to have a notable impact. Additionally, both trait anxiety and state anxiety were associated with diminished shifting efficiency, though their effectiveness remained unaffected. Importantly, no correlation was observed between either trait or state anxiety and updating capabilities.

These results align with the predictions of ACT that posit a significant relationship between anxiety and cognitive functioning. However, a distinction was noted: trait anxiety uniquely influenced both efficiency and effectiveness in inhibition tasks contrasting with the findings in [56]. Their study revealed that participants with high trait anxiety exhibited impaired processing efficiency across all three cognitive tasks analyzed, while their effectiveness was not compromised.

ACT elucidates the specific functions of the central executive most adversely impacted by anxiety, emphasizing the 'inhibition' and 'shifting' functions [32]. The inhibition function requires the exertion of attentional control to suppress interference from irrelevant stimuli (termed negative control), whereas the shifting function involves reallocating attention to remain focused on relevant stimuli (known as positive control). It is the disruption of these attentional control processes (particularly inhibition and shifting) that is believed to disturb the equilibrium between goal-directed and stimulus-driven attentional systems [13].

11.4 Attention Control Components

ACT forecasts trait anxiety to hinder three key executive functions: the inhibition function that involves using attentional control to suppress attention to task-irrelevant stimuli and to inhibit strong but incorrect responses; the shifting function that refers to the ability to redirect attention in response to changing requirements both within and between tasks; and the updating function that pertains to the processes of updating and monitoring information within working memory. We explain these three components throughout this section.

11.4.1 Inhibition

The primary prediction of ACT, which has undergone extensive empirical examination, is that anxiety disrupts the ability to effectively suppress task-irrelevant information. Numerous studies have supported this prediction by utilizing threat distractors. However, more recent research has indicated that anxiety may also impair the inhibition of task-irrelevant material even when no threat is present. Behaviorally, this effect is particularly evident in the antisaccade task, where individuals with high anxiety levels are slower to initiate eye movements away from an abrupt peripheral target compared to their low-anxiety counterparts. In contrast, they do not exhibit slower responses on prosaccade trials, where they simply look toward the target, thereby minimizing the need for inhibitory processes. The longer latency in antisaccade tasks suggests a diminished ability to inhibit attention to the salient item, highlighting a decrease in top-down control over reflexive processes, as predicted by ACT.

Beyond the antisaccade paradigm, reduced inhibition associated with anxiety has also been demonstrated through increased attentional capture by irrelevant stimuli and heightened interference from task-irrelevant distractors in response competition scenarios.

Anxiety seems to disrupt inhibition, but the neural mechanisms behind this effect are less clear. It might relate to difficulties in detecting conflict or to a diminished capacity for active inhibition [36, 37]. For instance, in [15], highly anxious individuals were identified to exhibit reduced dorsolateral prefrontal cortex (DLPFC) activity when faced with response-competing letter distractors in a visual search task, suggesting a decreased ability to inhibit irrelevant information [34].

Additionally, lower fronto-central negativity is observed in [29] before the target in high-anxiety individuals during an antisaccade task, correlating with longer eye-movement latencies in these participants; This indicates a poor recruitment of top-down resources in those with anxiety [15].

Conversely, other studies indicate that anxiety is linked to increased neural processing related to inhibition. For example, in a sustained attention task, highly anxious participants displayed a larger N2 component response without increased behavioral errors in [31] suggesting enhanced inhibitory processing. Similarly, greater EEG desynchronization in high-anxiety individuals during a stop-signal task is observed in [38], indicating more effort and resource allocation to inhibit motor responses [31–38].

Lastly, a recent fMRI study; [30], revealed highly anxious participants showing increased DLPFC activity during incongruent trials in a Stroop task, contrasting with findings in [15] and suggesting greater cognitive resource recruitment [15–30].

11.4.2 Shifting

ACT posits that anxiety is associated with a reduced capacity to flexibly shift attention between relevant tasks. This assertion is supported by evidence from studies using the Wisconsin Card Sorting Task, which demonstrates that anxiety leads to increased errors and longer reaction times. Research conducted in [3] further examined the impact of anxiety on attentional shifting, revealing that highly anxious participants were significantly slower in a task-switching paradigm, with their performance delays worsening as task complexity increased. In contrast, individuals with low anxiety demonstrated consistent performance across different conditions. Similarly, findings in [29] indicated that the benefits of switching in antisaccade tasks were observed only among low-anxiety participants.

However, there is a scarcity of studies investigating neural correlations associated with shifting functions in individuals experiencing anxiety. One notable exception is conducted in [29] as a mixed antisaccade task with both prosaccade and antisaccade trials, where the trials were indicated by the color of a fixation cross. The authors varied the inter-stimulus interval (ISI) between the fixation cross and the target. Their results revealed that at shorter ISIs, highly anxious individuals exhibited slower responses on antisaccade trials, while at longer ISIs, there were no significant latency differences between the groups. Event-related potentials (ERPs) indicated that anxious participants displayed increased slow wave negativity at longer ISIs, suggesting that they were exerting greater cognitive effort and resource allocation to adapt to changing task demands. These findings imply that anxious individuals may employ more compensatory strategies to sustain performance levels comparable to their less anxious counterparts, particularly at longer ISIs where preparatory time is available.

11.4.3 Updating

The final aspect of working memory affected in individuals with anxiety, as outlined by Acceptance and Commitment Therapy, is the process of updating information. This process entails monitoring and revising the contents of working memory. Behavioral findings related to this area have been somewhat variable, indicating that anxiety may interfere with updating mechanisms primarily in high-stress situations. While research in neuroscience remains limited, there have been advancements in understanding how anxiety influences error responses during cognitive tasks, specifically through the lens of conflict monitoring, which is closely tied to working memory functions.

Notably, research consistently shows heightened activity in the anterior cingulate cortex (ACC) among individuals with high anxiety levels. For instance, event-related potentials (ERP) are used in [57] to demonstrate that increased anxiety correlates with a more pronounced error-related negativity (ERN) following mistakes in a Stroop task, with the amplitude of the ERN reflecting ACC involvement. This enhanced ERN has been corroborated by several more recent studies. Additionally, the N400 component, which is also linked to conflict monitoring, represents ACC activity and is found to be more pronounced in highly anxious participants during tasks requiring selective attention.

Despite the lack of observable behavioral performance differences between anxious and non-anxious groups, those with anxiety show greater neural activation in conflict monitoring processes post-error, suggesting they engage more cognitive resources to achieve comparable performance levels. Nonetheless, there is a notable absence of comprehensive neuroscientific investigations exploring monitoring and updating processes within working memory specifically related to anxiety, such as in n-back tasks where participants are required to retain and continuously refresh information.

11.5 Discussion

The intricate relationship between anxiety disorders and attentional control is elucidated through various mechanisms, prominently highlighted in recent cognitive psychology and neuroscience research [1–57]. In this work, we emphasized that anxiety, both as a transient state and a stable trait, can significantly impair cognitive functions by altering how attentional resources are allocated, particularly under conditions of perceived threat.

A notable insight from this work is that the conceptual framing of anxiety is not merely an emotional state but a complex cognitive phenomenon that influences attentional processes on multiple levels. The Attention Control Theory provides a robust framework for understanding how anxiety compromises attentional control by diminishing the influence of top-down processes and enhancing the bottom-up processing of potentially threatening stimuli. This imbalance may explain the documented increases in reaction times and the enhanced processing of irrelevant distractors observed in anxious individuals.

The dual-process perspective on attentional bias—initial vigilance followed by potential avoidance or difficulty in disengaging—offers a nuanced understanding of how attentional systems operate under anxiety. This aligns with the notion that while attentional biases toward threat are common in anxiety, the degree and manifestation of this bias are dependent on the dynamic interplay of cognitive control systems.

Despite pervasive evidence supporting the attentional bias and cognitive control deficits in anxiety, the exact causal mechanisms remain partially understood. Although consistent findings exist on how anxiety impacts tasks requiring cognitive control and attentional shifts, there is a need for further exploration of whether these impacts are uniform across different cognitive loads and whether they vary depending on the specific executive function involved; inhibition, shifting, or updating.

Moreover, the review in this work underscores that while deficits in attentional control are apparent, they may not always translate to decreased performance effectiveness. High-functioning anxious individuals can maintain accuracy by exerting additional cognitive effort, highlighting a hidden compensatory cost. This observation suggests that future research should focus not only on performance outcomes but also on the cognitive and emotional resources expended to achieve those outcomes.

The bidirectional influence of anxiety and attentional control posited by ACT suggests potential developmental implications, where poor attentional control might exacerbate anxiety symptoms over time and vice versa. This relationship underscores the importance of early interventions targeting attentional processes to potentially mitigate the progression of anxiety disorders.

In conclusion, although the evidence for ACT's predictions regarding the distinct executive functions remains mixed, this review reinforces the notion that anxiety substantially affects processing efficiency across various cognitive tasks. Future investigations should aim to disentangle the complexities of attentional control mechanisms under anxiety, examining the differential impacts on task effectiveness and efficiency across varied cognitive demands. Additionally, different disease categories impact on the associated outcomes can be investigated in order to investigate if different diseases can be impacting the corresponding outcomes.

References

[1] Pine, D. S., P. Cohen, D. Gurley, J. Brook and Y. Ma. 1998. The risk for early-adulthood anxiety and depressive disorders in adolescents with anxiety and depressive disorders. Archives of General Psychiatry 55(1): 56–64.

[2] Pine, D. S. 2007. Research review: a neuroscience framework for pediatric anxiety disorders. J. Child Psychol Psychiatry 48(7): 631–48.

[3] Derakshan, N. and M. W. Eysenck. 2009. Anxiety, processing efficiency, and cognitive performance: New developments from attentional control theory. European Psychologist 14: 168–176.

[4] Bar-Haim, Y., D. Lamy, L. Pergamin, M. J. Bakermans-Kranenburg and M. H. Van Ijzendoorn. 2007. Threat-related attentional bias in anxious and nonanxious individuals: a meta-analytic study. Psychological Bulletin 133(1): 1.

[5] Dudeney, J., L. Sharpe and C. Hunt. 2015. Attentional bias towards threatening stimuli in children with anxiety: A meta-analysis. Clinical Psychology Review 40: 66–75.

[6] Mogg, K. and B. P. Bradley. 1998. A cognitive-motivational analysis of anxiety. Behaviour Research and Therapy 36: 809–848.

[7] Williams, J. M. G., A. Mathews and C. MacLeod. 1996. The emotional Stroop task and psychopathology. Psychological Bulletin 120: 3–24.
[8] Amir, N., E. B. Foa and M. E. Coles. 1998. Automatic activation and strategic avoidance of threat-relevant information in social phobia. Journal of Abnormal Psychology 107: 285–290.
[9] Mogg, K., B. P. Bradley, J. De Bono and M. Painter. 1997. Time course of attentional bias for threat information in non-clinical anxiety. Behaviour Research and Therapy 35: 297–303.
[10] Williams, J. M. G., F. N. Watts, C. MacLeod and A. Mathews. 1988. Cognitive Psychology and Emotional Disorders. John Wiley and Sons.
[11] Williams, J. M. G., F. N. Watts, C. MacLeod and A. Mathews. 1997. Cognitive Psychology and Emotional Disorders (2nd ed.). Chichester, England: Wiley.
[12] Van Ryckeghem, D., M. Noel, L. Sharpe, T. Pincus and S. Van Damme. 2019. Cognitive biases in pain: An integrated functional-contextual framework. Pain 160: 1489–1493.
[13] Eysenck, M. W., N. Derakshan, R. Santos and M. G. Calvo. 2007. Anxiety and cognitive performance: Attentional control theory. Emotion 7: 336–353.
[14] Eysenck, M. W. and N. Derakshan. 2011. New perspectives in attentional control theory. Personality and Individual Differences 50(7): 955–960.
[15] Bishop, S. J. 2009. Trait anxiety and impoverished prefrontal control of attention. Nature Neuroscience 12: 92–98.
[16] Fox, E. 1993. Attentional bias in anxiety: selective or not? Behaviour Research and Therapy 31: 487–493.
[17] Pacheco-Unguetti, A. P., A. Acosta, A. Callejas and J. Lupiánez. 2010. Attention and anxiety: different attentional functioning under state and trait anxiety. Psychological Science 21: 298–304.
[18] Eysenck, M. W. 1996. Anxiety, processing efficiency theory and performance. pp. 91104. *In*: Battman, W. and S. Dutke (eds.). Processes of the Molar Regulation of Behaviour. Lengerich: Pabst Science.
[19] Neiss, R. 1988. Reconceptualizing arousal: Psychobiological states in motor performance. Psychological Bulletin 103: 34566.
[20] Langner, R. and S. Eickhoff. 2013. Sustaining attention to simple tasks: A meta-analytic review of the neural mechanisms of vigilant attention. Psychological Bulletin 139: 870–900. doi:10.1037/a0030694.
[21] Cornwell, B. R., J. M. P. Baas, L. Johnson, T. Holroyd, F. W. Carver, S. Lissek et al. 2007. Neural responses to auditory stimulus deviance under threat of electric shock revealed by spatially-filtered magnetoencephalography. Neuroimage 37: 282–289. doi:10.1016/j.neuroimage.2007.04.055.
[22] Liebert, R. M. and L. W. Morris. 1967. Cognitive and emotional components of test anxiety: a distinction and some initial data. Psychological Reports 20: 975978.
[23] Kahneman, D. 1973. Attention and Effort. Englewood Cliffs, NJ: Prentice-Hall.
[24] Baumeister, R. F. 1984. Choking under pressure: Self-consciousness and paradoxical effects of incentives on skillful performance. Journal of Personality and Social Psychology 46: 610620.
[25] Eysenck, M. W., J. S. Moser, N. Derakshan, P. Hepsomali and P. Allen. 2022. A neurocognitive account of attentional control theory: how does trait anxiety affect the brain's attentional networks? Cognition and Emotion 37(2): 220–237.
[26] Corbetta, M. and G. L. Shulman. 2002. Control of goal-directed and stimulus-driven attention in the brain. Nature Reviews Neuroscience 3: 201–215.
[27] Miyake, A., N. P. Friedman, M. J. Emerson, A. H. Witzki, A. Howerter et al. 2000. The unity and diversity of executive functions and their contributions to complex 'frontal lobe' tasks: A latent variable analysis. Cognitive Psychology 41: 49100.
[28] Berggren, N. and N. Derakshan. 2013. Attentional control deficits in trait anxiety: Why you see them and why you don't. Biological Psychology 92(3): 440–446.
[29] Ansari, T. L. and N. Derakshan. 2011a. The neural correlates of cognitive effort in anxiety: effects on processing efficiency. Biological Psychology 86(3): 337–348.
[30] Basten, U., C. Stelzel and C. J. Fiebach. 2011. Trait anxiety modulates the neural efficiency of inhibitory control. Journal of Cognitive Neuroscience 23(10): 3132–3145.
[31] Righi, S., L. Mecacci and M. P. Viggiano. 2009. Anxiety, cognitive self-evaluation and performance: ERP correlates. Journal of Anxiety Disorders 23(8): 1132–1138.

[32] Miyake, A., N. P. Friedman, M. J. Emerson, A. H. Witzki, A. Howerter and T. D. Wager. 2000. The unity and diversity of executive functions and their contributions to complex frontal lobe tasks: a latent variable analysis. Cognitive Psychology 41(1): 49–100.
[33] Cisler, J. M. and E. H. W. Koster. 2010. Mechanisms of attentional biases towards threat in anxiety disorders: an integrative review. Clinical Psychology Review 30: 203–216.
[34] Moser, J. S., M. W. Becker and T. P. Moran. 2012. Enhanced attentional capture in trait anxiety. Emotion 12(2): 213.
[35] Poy, R., M. D. Eixarch and C. Ávila. 2004. On the relationship between attention and personality: covert visual orienting of attention in anxiety and impulsivity. Personality and Individual Differences 36: 1471–1481.
[36] Botvinick, M. M., T. S. Braver, D. M. Barch, C. S. Carter and J. D. Cohen. 2001. Conflict monitoring and cognitive control. Psychological Review 108(3): 624.
[37] Aarts, K. and G. Pourtois. 2010. Anxiety not only increases, but also alters early error-monitoring functions. Cognitive, Affective, and Behavioral Neuroscience 10: 479–492.
[38] Savostyanov, A. N, A. C. Tsai, M. Liou, E. A. Levin, J. D. Lee, A. V. Yurganov et al. 2009. EEG correlates of trait anxiety in the stop-signal paradigm. Neuroscience Letters 449: 112–116.
[39] Goodwin, A. H. and K. J. Sher. 1992. Deficits in set-shifting ability in non-clinical compulsive checkers. Journal of Psychopathology and Behavioral Assessment 14: 81–92.
[40] Caselli, R. J., E. M. Reiman, J. G. Hentz, D. Osbourne and G. E. Alexander. 2004. A distinctive interaction between chronic anxiety and problem solving in asymptomatic APOE e4 homozygotes. Journal of Neuropsychiatry and Clinical Neurosciences 16: 320–329. https://doi.org/10.1176/appi.neuropsych. 16.3.320.
[41] Jennings, J. R. and M. W. van der Molen, 2005. Preparation for speeded action as a psychophysiological concept. Psychological Bulletin 131: 434–459.
[42] Sohn, M. H., M. V. Albert, K. Jung, C. S. Carter and J. R. Anderson. 2007. Anticipation of conflict monitoring in the anterior cingulate cortex and the prefrontal cortex. Proceedings of the National Academy of Sciences of the United States of America 104: 10330–10334.
[43] Olvet, D. M. and G. Haycak. 2009. The effect of trial-to-trial feedback on the error-related negativity and its relationship with anxiety. Cognitive, Affective, and Behavioral Neuroscience 9: 427–433.
[44] Aarts, K. and G. Pourtois. 2010. Anxiety not only increases, but also alters early error-monitoring functions. Cognitive, Affective, and Behavioral Neuroscience 10: 479–492.
[45] Badzakova-Trajkov, G., K. J. Barnett, K. E. Waldie and I. J. Kirk. 2009. An ERP investigation of the Stroop task: the role of the cingulated in attentional allocation and conflict resolution. Brain Research 1253: 139–148.
[46] Badzakova-Trajkov, G., K. J. Barnett, K. E. Waldie and I. J. Kirk. 2009. An ERP investigation of the Stroop task: the role of the cingulated in attentional allocation and conflict resolution. Brain Research 1253: 139–148.
[47] Osinsky, R., N. Alexander, H. Gebhardt and J. Hennig, 2010. Trait anxiety and dynamic adjustments in conflict processing. Cognitive, Affective, and Behavioral Neuroscience 10: 372–381.
[48] Berggren, N. and N. Derakshan. 2013. Attentional control deficits in trait anxiety: why you see them and why you don't. Biol Psychol. 92(3): 440–6.
[49] Shi, R., L. Sharpe and M. Abbott. 2019. A meta-analysis of the relationship between anxiety and attentional control. Clin. Psychol. Rev. 72: 101754.
[50] Shechner, T., J. C. Britton, K. Pérez-Edgar, Y. Bar-Haim, M. Ernst, N. A. Fox et al. 2012. Attention biases, anxiety, and development: toward or away from threats or rewards? Depress Anxiety 29(4): 282–94.
[51] Grillon, C., O. J. Robinson, A. Mathur and M. Ernst. 2015. Effect of attention control on sustained attention during induced anxiety. Cognition and Emotion 30(4): 700–712. https://doi.org/10.1080/02699931.2015.1024614.
[52] Sari, B. A., E. H. Koster, G. Pourtois and N. Derakshan. 2016. Training working memory to improve attentional control in anxiety: A proof-of-principle study using behavioral and electrophysiological measures. Biol Psychol. 121(Pt B): 203–212.

[53] Wilson, M. 2008. From processing efficiency to attentional control: a mechanistic account of the anxiety–performance relationship. International Review of Sport and Exercise Psychology 1(2): 184–201. https://doi.org/10.1080/17509840802400787.
[54] Altan-Atalay, A. 2018. Moderator Role of Looming Cognitive Style (LCS) in the relationship between attentional control and anxiety: difference between shifting and focusing dimensions. The Journal of Psychology 152(7): 445–455. https://doi.org/10.1080/00223980.2018.1466775.
[55] Edwards, E. J., M. S. Edwards and M. Lyvers. 2015. Cognitive trait anxiety, situational stress, and mental effort predict shifting efficiency: Implications for attentional control theory. Emotion 15(3): 350.
[56] Wong, I., D. Mahar, K. Titchener and J. Freeman. 2013. The impact of anxiety on processing efficiency: Implications for the attentional control theory. The Open Behavioral Science Journal 7(6): 7–15.
[57] Hajcak, G., N. McDonald and R. F. Simons. 2003. To err is autonomic: Error-related brain potentials, ANS activity, and post-error compensatory behavior. Psychophysiology 40(6): 895–903.

CHAPTER 12

Brain and Body Perception of Beauty

Emre Tokgoz,[1,*] *Percin Karakol,*[2] *Bilge Kaan Kilicoglu*[2] and *Sophia Marra*[3]

12.1 Introduction

For centuries, people have tended to be beautiful and participated in decision making of what may be beautiful. People have searched for the beauty of nature, on the constructions they made, and even within themselves. Such search and attempt for incorporating beauty into life today even includes artificial intelligence use for altering images to appear beautiful in social media platforms and information exchanges along with going under plastic surgeries. It is natural that each person's beauty perception changes along with the changes in environment, experiences, and ethnicity [4]. Utilization of all five senses in beauty decision making along with firing of neurons across the human body and activation of hormones form a very complex system of decision making judged by the brain. Even though some may believe the boundary of decision making on beauty is mostly spiritual and subjective, the brain has the final decisive factor.

Human beauty is frequently defined in terms of physical attributes that are perceived as attractive to potential mates [1]. In that case, defining the beauty of a person can depend on multiple factors and these factors are regulated according to their conscious ideas and their subconscious instincts. Evolutionary pressures are explained to be shaping human physical attributes and their neural processing

[1] School of Engineering and Technology, Farmingdale State College, New York, 11735 USA.

[2] Başakşehir Cam and Sakura City Hospital Plastic Reconstructive and Aesthetic Surgery, Istanbul, Turkiye.

[3] Frank H. Netter School of Medicine, Quinnipiac University, North Haven, CT, 06473, USA.

* Corresponding author: tokgoze@farmingdale.edu

DOI: 10.1201/9781003646044-12

to maximize reproductive fitness in [2]. To achieve such maximization, when a male or female sees a potential mate, the human brain starts to collect the data with the senses. There are at least three distinct neural networks employed with each comprising interconnected brain areas to assess facial attractiveness: one dedicated to recognition, another to interpretation, and the third one to evaluation. Key determinants of such judgment evaluated in the literature included age, health status, symmetry, averageness, facial and bodily proportions, as well as skin tone, use of cosmetics, and texture. These attributes are observed to be reliable indicators of reproductive fitness due to their inherent difficulty in being counterfeited [2]. Also in another study, scientists address that human attractiveness confers significant evolutionary advantages, as evidenced by heightened reproductive success, enhanced socioeconomic status, and extended lifespan [5].

The tradeoff between deciding what is beautiful based on personality traits and physical appearances is always one area of interest in research. One of the earlier review studies investigated the understanding of "what is beautiful is good" that was investigated in [54] around 1970s and determined that personality traits and attained more successful life outcomes are dominant aspects of decision making favoring attractive persons than unattractive ones [53]. The authors also concluded that "beauty is good" effect was moderate based on the observations.

The remainder of this work is organized as the next section focuses on the brain's perception of beauty and attractiveness, third section covering human beauty features and the associated observations. The fourth section covers facial attractiveness, while fifth section investigates bodily attractiveness. Female versus male attractiveness is covered in the sixth section with what follows it to be the methodology and conclusive remarks.

12.2 Brain's Perception of Beauty and Attractiveness

Neuroaesthetics is a relatively new field of research that studies how the brain interprets beauty and aesthetics [13]. With neuroimaging, it is shown that at least 3 areas of the brain (affective, cognitive, and sensorimotor) have roles in interpreting beauty. For example, when a person sees another person, the human brain can decide in milliseconds whether that person is beautiful or not. In that procedure, some of the brain loci attempts to fulfill specific roles. First, the sensory-motor cortex including the lateral fusiform gyrus (also called the fusiform face area) processes information and interprets incoming information and data. After this interpretation, the cognitive system (dorsolateral prefrontal cortex) integrates the data and presents the affective system for aesthetic evaluation. In another research, the affective system is shown to be composed of multiple areas of the cortical orbitofrontal cortex (that has a strong connection with the fusiform face area) and subcortical areas (such as nucleus accumbens and ventral striatum) [28].

Two areas of the brain activate when attractiveness of the faces is evaluated when they are visualized; Medial orbitofrontal cortex (mOFC) is active while attractive faces are viewed, and bilateral insular cortexes (that is also a part of the orbitofrontal cortex) are activated while viewing unattractive faces [17]. Another research conducted confirms attractive faces particular stimulation of mOFC and nucleus accumbens (reward and addiction center of the brain) while unattractive faces stimulating the right lateral OFC. Some researchers also found a relationship between the amygdala and aesthetic judgment but there is no direct correlation found between the amount of stimulation and beauty.

One of the areas studied in identification of attractiveness is associated with the associated neural correlations for facial beauty [29]. fMRI is used by the authors to identify empirical outcomes for addressing this research question based on the judgements of individuals' facial beauty explicitly. The neural network design components included ventral occipital, anterior insular, dorsal posterior parietal, inferior dorsolateral, and medial prefrontal cortices are identified to be correlated with the degree of facial attractiveness. Judging facial identity rather than attractiveness is identified to be activate the ventral occipital region that also remained to be responsive while judging facial beauty. Hence, a wide area of neural network involving perceptual, decision-making and reward circuits is facilitated during observations of facial beauty. Lateral occipital cortex and fusiform face area remained active during the perceptual response even during subjects not paying explicit attention to facial beauty.

The rewards center activation upon viewing attractive faces have been observed in several studies [19, 57] motivating sexual behavior along with same-sex alliance development [58, 59, 60] and being treated positively within a variety of settings [61]. The question of what it means to have an attractive face remains open with personal preferences and judgement, along with an increase in this question's scope that includes advanced technologies such as artificial intelligence, and surgical procedures such as cosmetic and plastic surgeries. For instance, in a study that investigated the effect of facial ratings including attractiveness, beauty, trustworthiness and familiarity on simulation monitoring (representing the belief that a stimulus was artificially generated), all 150 research participants believed a significant proportion of the images that were taken from real faces from the same database were fake (with high confidence) underlining the volatility of sense of reality [62]. The authors of the work also identified a high positive linear relationship in male participants for both attractiveness and beauty on simulation monitoring with males particularly emphasizing facial attractiveness to be a sign of reproductive potential when compared with females that had the tendency to value characteristics signaling resource acquisition capabilities [63, 64, 65].

A review of 37 research articles exhibiting psychological outcomes on patients that had plastic surgery indicated a general satisfaction of patients with the outcome of their plastic surgery procedures while some of the patients exhibiting psychological disturbances that are transient while some others experiencing long-lasting psychological disturbances [52]. In the study, poor psychosocial outcome-related factors are determined to include the following:

- Unrealistic outcomes expectations
- Previously experienced unsatisfactory cosmetic surgery
- Young age

- Being male
- Minimal deformity
- Issues based on relationships
- Experiences on anxiety, depression, or personality disorder.
- Body dysmorphic disorder

The authors concluded recommended a careful approach to cosmetic surgery decision making for individuals even though most people appear to be satisfied with the outcome of cosmetic surgical procedures as some people are not [52].

12.3 Human Beauty Features and Observation

A variety of features can be playing roles in determining human beauty. These criteria can change be changing due to factors such as an individual's aesthetic sense and gender, but some classical features can affect the perception of beauty. There are a variety of methodologies used for investigating and measuring human attractiveness with the major focus on the impact of facial attractiveness to be in one of the major areas of such investigations and, not surprisingly, fMRI studies to be the forefront approach for determining brain region activities. The other areas of investigations included body part-related observations when it comes to human attractiveness and the associated investigations that included but not limited to the following [33]:

- Age
- Health status
- Symmetry
- Facial proportions and symmetry
- Foot size
- Wrinkles
- Facial color gradient

There are additional external factors that could play a role in human attractiveness aside from the physical features that are based on selection of the individual and such factors are a result of choice that relies on self-modification that include but are not limited to the following:

- Wearing make-up
- Social status and influence
- Cultural ideals and norms
- Peer influence
- Familiarity with the individual
- Sexual dimorphism
- Body proportions

An individual's judgement of health is another factor driving the beauty concept and decision making. Appearance cues developed by individuals based on the history of the individual and the existing state of health are several such related decisions. For instance, coloring and darkening of sclera occurs due to poor health or age while whiter sclera is an associated with youngness that impact perception of attractiveness and health [37].

Eyes are also identified to be among the features of beauty selection. The role of amygdala in emotional processing of actions includes eye recognition [41] while superior temporal sulcus is used for eye movement tracking and identifying direction of views [40].

The level of attractiveness among genders based on the foot size is also researched in the literature [43, 44]; Small woman feet has been observed to be found more attractive by both men and women. A set of altered images have been shown to both men and women to decide preference of men feet as a means of attractiveness, and participants favored the natural proportion of men feet to height when compared to over- or under-sized feet.

12.4 Facial Attractiveness

The facial analysis provides some insights beyond basic demographic information, such as sex, age, and ethnicity. It also reveals nuanced social cues, including emotional state, trustworthiness, attractiveness, and intent [1]. The factors contributing to facial attractiveness and the extent to which these preferences are shaped by culture or biology have been the subject of scholarly inquiry for centuries [3]. In humans, the facial processing area is basically in the lateral fusiform gyrus (the fusiform face area), part of the occipitotemporal cortex. The occipitotemporal cortex also plays a role in the facial processing. Studies show that the evaluation of facial attractiveness occurs within a mere 100 milliseconds [6]. In that processing and evaluation part, there are multiple features; It can be classified as symmetry, averageness, and masculinity-femininity [2].

In many articles facial symmetry is observed to cause more attractiveness, and it is thought to stem from the symmetry showing gene quality and a healthy state. On the other hand, results about perfect symmetry were assessed as more attractive in [3] and less attractive in another [14]. These studies both used a qualitative assessment methodology that includes an adequate number of participants, symmetry-graded photographs, and computed evaluations for statistics.

Facial attractiveness is determined to be one of the major reasons for attractiveness and the attractiveness of individuals is judged by some people based on the age [34] that is correlated with the attractiveness level based on age by some others [35]. Additional features to age are included in the research such as skin and body types, reproductive potential, and desirability as a partner to observe individuals' selection of partners. Physiological factors in addition to the economic factors play roles in decision making in societal settings in a way that men can be more desirable by women due to accumulating resources while men's perception of women attractiveness can be driven by the menopause and fertility changes due to aging as a boundary condition [36].

Facial beauty preferences can be depending on a variety of features that is measured by using the concept of "average facial features" representing the typical characteristics and proportions that are found in a population [39]. When the averageness of the face is evaluated, several theorist researchers suggested the average traits to be indicative of developmental stability, and the capacity to resist stress during development therefore that averageness could indicate the mate quality. Studies show that having a more average face increases attraction [15]. Despite average faces to be found more attractive in several studies; a study showed an accepted beautiful face to be not close to this average [16]. This may be an indicator that perceiving beauty of each person is not behaving in a parallel manner to average attractivity but rather an indicator of each individual's subconscious appeal for averageness perceived.

The masculinity and femininity part of the facial evaluation is about hormonal regulation that affects humans from intrauterine part to end of the life. It has an essential role in beauty for the characteristics of the body and face. Numerous studies have demonstrated that men tend to find feminine features more appealing in female faces, while women often prefer masculine features in male faces [22, 23]. However, such a perception is not without limits; Conversely, some studies have failed to find a correlation between facial femininity or masculinity and attractiveness [24]. The variability in beautiful face decision making for everyone can be different and such decision making can be due to several different reasons.

One other factor in the facial beauty evaluation of the mind is driven by wearing makeup. A study showed women wearing makeup increasing their eye sizes [45] and such a change is identified to have the strongest impact on both men's and women's judgement of attractiveness [46]. In the case of general cosmetic modifications to the face, individuals' skin tone changes and reduced age appearance are two of the factors observed to be impactful on attractiveness to women that also impacted the perception of beauty [47, 48]. A uniformly distributed skin color is identified to be in correlation with the healthy appearance and increased attractiveness [49]. In such a regard, the skin color darkens as an individual ages causing a reduction on the color contrast between the eyes, hair, and facial skin [50]. Aging and sun exposure are two of the factors in wrinkling of skin that are used in age judgement among other factors, and perception of skin wrinkling is associated with loss of fertility as well [51].

Neuroscientific reasoning for remembering attractive faces better than others is investigated in [56] empirically using female participant data and event-related fMRI based on their rating of male faces' attractiveness. The major fMRI findings included the following:

- A linear increase of activity within the right orbitofrontal cortex is observed based on attractiveness ratings.
- Left hippocampus activity increased as a function of subsequent memory.
- Attractive faces caused stronger functional connectivity between the orbitofrontal and hippocampal regions.

Overall, a better memory for attractive faces is observed to reflect much better interactions between reward and successful memory encoding by activating the corresponding orbitofrontal cortex and hippocampus regions respectively. During the experiments, attractive faces were remembered better than other faces as expected.

12.5 Anatomical Attractiveness

After studying in Italy, German painter and mathematician Albrecht Dürer introduced the laws of proportion to Northern Europe. In his 1582 treatise, *De Symmetria*, he outlined a system of ideal human proportions. He simplified the body parts to the simple forms that can be measured and stated that the body proportion makes the body with a balanced, organic wholeness. The features that become attractive mostly do not disrupt the balanced wholeness such as symmetry, slimness, and height [42].

The notion of body symmetry is similar to facial symmetry. It determines genetic efficiency and well-being and therefore appears as a reason for attraction for both genders. The research revealed a significant positive correlation between the degree of bilateral symmetry and the quantity and quality of sperm in ejaculate, as measured by sperm count and motility which provides evidence for a potential positive relationship between the degree of bilateral symmetry and male reproductive potential [21].

Obesity is identified to decrease attractiveness for both genders in [2]. The ideal measurements for slimness are known as Body Mass Index (BMI) and Waist-to-Hip Ratio (WHR). A lower BMI is more correlated with health and genetic well-being that is determined to be preferable. On the other hand, WHR is an indicator for fat accumulation which is regulated by sex hormones. Estrogen has a significant role in storing fat in the hips and thighs and decreasing the fat storage in internal organs that can give the hourglass shape to a woman's body. A low WHR shows a higher level of estrogen and health, as well as higher reproductive ability. Also, height is another variable to bodily attractivity. It is reported that tall stature particularly increases attractiveness in men [26].

The evaluation of symmetry of body features is another factor playing a role in humans' decision making of beauty. This evaluation of symmetry is determined to correlate with the health position of the individual; Facial and anthropometric traits were investigated in [38] for deviations from perfect bilateral symmetry, and empirical results were attained based on measurements by using a range of physiological variables. The results indicated poor inter-correlations between the fluctuating asymmetries measured at different sites while no specific patterns of consistent significant correlations were attained between fluctuating asymmetry across different physiological variables over all traits.

Facial appearances and nude bodies of people are evaluated by 20 participants in [20] by using 3T-fMRI based on extremely beautiful, extremely ugly, and indifferent categorization. The findings of the research are reported based on different brain regional area activations:

- Highest activation of stimuli for both very ugly and very beautiful in basal occipial areas. Among such regions, mOFC, and powterior and medial cingulate gyrus were also activated.

- Esthetic judgements on ugliness based on a linear relationship activity between anterior cingulate cortex and caudate and nucleus accumbens as a part of lowest extreme for beauty judgement.

The authors concluded their findings on neural activity activation to be based on comparison of observed individuals' comparison of others to themselves. In addition, the faces and bodies evaluation of the participants based on the observed biological and social stimulation to be related to each other.

12.6 Female vs. Male Attractiveness

Attractiveness across genders has been studied extensively based on a variety of human body features. Sexual dimorphism is explained to be differences in physical features based on gender in [10], and such features are determined by sex hormones. While estrogen causes feminization, testosterone causes masculinization, and these play roles in attractiveness identification in the opposite gender. Such selections also support evolutionary biology due to hormones' impacting the looks of individuals with evaluation of women to have a feminine body and facial features, and a man to be more masculinized. Traits that deemed opposite sex attractiveness by the opposite sex are often associated with fertility, health, and youth—or, more broadly, genetic quality and the potential to be a good partner or parent. This suggests that humans subconsciously select partners in a way that optimizes their chances of reproduction and gene transmission which has a direct connection with attractivity [1]. In this context, there was a need for facial and bodily attractiveness to evaluate mating selection choices and human decisions with not only the conscious part but also the subconscious part of the brain. On the other hand, facial and bodily attractiveness have different effects on gender. Men, but not women, is mentioned to have a context-dependent adaptive tendency to prioritize facial cues in long-term mating relationships while shifting their focus to bodily cues in short-term mating situations [11].

The typical brain-based evaluation of attractiveness relied on fMRI studies regardless of human or other attractiveness perceptions. The role of brain in determination of human beauty is determined to have similar brain regional activities. For instance, an fMRI study was conducted on brain's perception of beaty based on the musical and painting pictures to investigate brain's regional activity correlation among 21 persons [7]. The empirical measurement relied on a 9-point scale with the enumeration relying on 1 being the least beautiful. The researchers classified the results into three categories that consisted of stimuli selections ugly, indifferent, and beautiful. Among the stimuli activity in various regions of the brain, the medial orbito-frontal cortex (m-OFC) in the cortical area is determined to be active during the evaluation of visual and musical activities. The measurements of three-level intensity were identified to be proportional to the strength of the declared intensity of the experience of the individuals with beauty. Similarly, the role of brain during facial attraction decision-making has also been studied extensively in research literature. Brain's selection of facial attributes to decide whether a face is beautiful or not was determined to follow a linear response mechanism within orbito-frontal cortex (OFC); a region that plays a role in rewarding processes as we pointed out previously. On the contrary of this linear approach, the involvement of amygdala in the facial attractiveness is questioned empirically to understand its role in the facial attractiveness determination and rewarding process; The fMRI results on human subjects were collected in [5, 29] based on the rating of facial attractiveness and age. Independent of the task assigned, right amygdala is identified to be responsive and involved during the decision-making process of highly unattractive and attractive individuals when compared to middle ranked faces. A variability of linear and non-linear patterns of behavior was observed in OFC across different regions of it. In such determination, there are furthermore details on the decision-making of attractiveness that includes the symmetric nature of the face [30] and age [32], and such decisions psychologically impact the decision maker. It has been observed in [32] that attractiveness was the primary decision-making component during the mate selection process of men, and age was one of the factors in such determination. Overall, a variety of empirical observations have been made, and a meta-analytic of the research literature revealed distinct neural specificities for visual art and facial beauty that lacked the evidence for the common neural basis of visual beauty [31].

The physiological and psychological connection between the sensitivity to beauty decisions is clear in many ways; The major question is related to determination of the factors that play significant roles in such decisions. Brain activates neural networks associated with the rewarding processes that relate to regions such as orbitofrontal cortex, nucleus accumbens, and ventral striatum with beauty may be holding privileged attention of brain even when attention is paid elsewhere. The beauty may be taking over the attention consciously or unconsciously with the contextual motivation also playing a significant role in decision-making processes that relates to relationship status, sociosexual orientation, gender, and level of visual acuity of the individual [27].

12.7 Social Psychology Methodologies

Initiated by "what is beautiful is good" approach to beauty in 1972 ([54]), methodological observations on the concept of investigating and searching for the meaning of beauty of human being is still continuing for researchers with the complexity of such investigations increasing as new technologies are introduced. As a part of social psychological research, there have been various findings and approaches utilized:

- An attractive target leads perceivers to make strong inferences of personality goodness with exceptions existing to this phenomenon.
- Social competence and interpersonal ease are suggested to be strong measures in [66] that are in agreement with the beauty-is good effect
- Physical attractiveness is investigated in [68] with a conclusion similar to [67] with the roles of coail vitality and extroversion.
- Perception of individuals on attractive people's vainness and egoistical behavior is argued in [69] whether such beauty depends on one's self-centered behavior.

- Physical attractiveness stereotyping based on the implicit personality theory with researchers investigating relevant considerations. For instance, the linkage among the group members with group membership as one of the personal attributes inferentially associated with other attributes such as implicit theory [70].

Noting that there are many more approaches. The above-mentioned approaches are some of the fundamental findings in the area of interest for attempting to understand what is beautiful with the corresponding attraction investigation in social psychology.

12.8 Conclusion

A multitude of factors is evaluated by the human brain upon surveying such factors that relate to physiological and psychological settings. The relevant brain-based fMRI results outline the role of different regions along with the associated hormonal activity statuses. Even though some of the articles supported the beauty to be a dynamic collection of physical attributes such as averageness, facial symmetry, and skin uniformity while it can rather be a dynamic construct of continuously evolvement to responses that actively update by both conscious and subconscious influences, most articles show the body perception of beauty imposing a fixed stimuli and evaluations on the human brain. The relevant subconscious evaluations create an ideal structure for potential partner selection and allow humans to select the optimal beauty on individual basis. Looking at the fMRI studies, the specific location of brain activation can be based on a multitude of different considerations and perceptions. Given the psychological factors, the definition of optimal beauty is based on the selection criteria of the individual that is likely to be an outcome of collective life experiences and perspectives and the associated psychological factors along with potential diseases that are developed throughout the lifespan. The complexity of such system is likely to incorporate the belief mechanism of the individual along with the nervous system activities that depend on the life conditions of the individual. Overall, the results presented in the research literature are limited in their permitted investigations and the duration of the research timeframe.

12.9 Discussion

There are studies in literature that have controversial outcomes in the ways that the individuals perceive beauty that can rely on objective ways of identification such as ERP and fMRI results as well as subjective ways that relied on survey outcomes. Regardless of the method of collection, noting the variability of human brain among individuals, there are multiple factors impacting the ways beauty is perceived by the brain with potential observational biases.

A comprehensive study may be carried out to observe individuals' perception of beaty by looking at a variety of factors including the following:

- Family history (including associated illnesses and living conditions),
- Social life history

- Childhood experiences
- Individual's illnesses (both physiological and psychological)
- Traumatic events related to the context of observation on beauty,
- History of the individual on the perception of beauty including the changes occurred in such perception (if applicable),
- fMRI study
- Additional other factors outlined in this study including attraction to different physical and psychological factors.

Psychological factors such as body dysmorphic disorder (that is associated with experiencing significant stress due to excessive worries on physical appearance or a minor physical anomaly) impact the nervous system that works with all the systems in the body with these systems direct communication with the brain; hence, human body's perception of beauty is difficult to properly evaluate as the psychological conditions stem from an intricate interplay of impairments across various brain networks [8]. Other factors that can affect the relevant studies' validity include but not limit to the number of participants in the study, the sociocultural state, and age. Even though some of the studies attempted to optimize such factors, for a proper evaluation, a much wider scope of consideration is necessary for an adequate and extensive evaluation.

References

[1] Esel, E. and G. P. Esel. 2017. Güzellik Algılamasının Nörobiyolojisi ve Evrimsel Temelleri. Dusunen Adam: Journal of Psychiatry and Neurological Sciences 30(4).

[2] Yarosh, D. B. 2019. Perception and deception: human beauty and the brain. Behavioral Sciences 9(4): 34.

[3] Rhodes, G., F. Proffitt, J. M. Grady and A. Sumich. 1998. Facial symmetry and the perception of beauty. Psychonomic Bulletin and Review 5(4): 659–669.

[4] Dimitrov, D. and G. Kroumpouzos. 2023. Beauty perception: A historical and contemporary review. Clinics in Dermatology 41(1): 33–40.

[5] Winston, J. S., J. O'Doherty, J. M. Kilner, D. I. Perrett and R. J. Dolan. 2007. Brain systems for assessing facial attractiveness. Neuropsychologia 45(1): 195–206.

[6] Willis, J. and A. Todorov. 2006. First impressions: Making up your mind after a 100-ms exposure to a face. Psychological Science 17(7): 592–598.

[7] Ishizu, T. and S. Zeki. 2011. Toward a brain-based theory of beauty. PloS One 6(7): e21852.

[8] Feusner, J. D., J. Yaryura-Tobias and S. Saxena. 2008. The pathophysiology of body dysmorphic disorder. Body Image 5(1): 3–12.

[9] Sisti, A., N. Aryan and P. Sadeghi. 2021. What is beauty? Aesthetic Plastic Surgery 45(5): 2163–2176.

[10] Chatterjee, A. 2014. The aesthetic brain: How we evolved to desire beauty and enjoy art. OUP Us.

[11] Confer, J. C., C. Perilloux and D. M. Buss. 2010. More than just a pretty face: Men's priority shifts toward bodily attractiveness in short-term versus long-term mating contexts. Evolution and Human Behavior 31(5): 348–353.

[12] Merino, M., J. F. Tornero-Aguilera, A. Rubio-Zarapuz, C. V. Villanueva-Tobaldo, A. Martín-Rodríguez and V. J. Clemente-Suárez. 2024. Body perceptions and psychological well-being: a review of the impact of social media and physical measurements on self-esteem and mental health with a focus on body image satisfaction and its relationship with cultural and gender factors. In Healthcare (12: 1396). MDPI.

[13] Nakamura, K. and H. Kawabata. 2016. Beauty measured and manipulated by the brain: the psychophysiology of beauty. Japanese Journal of Physiological Psychology and Psychophysiology 34(1): 9–26. doi:10.5674/jp.1602ci.

[14] Swaddle, J. P. and I. C. Cuthill. 1995. Asymmetry and human facial attractiveness: symmetry may not always be beautiful. Proceedings of the Royal Society of London. Series B: Biological Sciences 261(1360): 111–116.

[15] Gunes, H. 2011. A survey of perception and computation of human beauty. In Proceedings of the 2011 Joint ACM Workshop on Human Gesture and Behavior Understanding. 19–24.

[16] Alley, T. R. and M. R. Cunningham. 1991. Article commentary: averaged faces are attractive, but very attractive faces are not average. Psychological Science 2(2): 123–125.

[17] O'Doherty, J., J. Winston, H. Critchley, D. Perrett, D. M. Burt and R. J. Dolan. 2003. Beauty in a smile: the role of medial orbitofrontal cortex in facial attractiveness. Neuropsychologia 41(2): 147–155.

[18] Cloutier, J., T. F. Heatherton, P. J. Whalen and W. M. Kelley. 2008. Are attractive people rewarding? Sex differences in the neural substrates of facial attractiveness. Journal of Cognitive Neuroscience 20(6): 941–951.

[19] Rhodes, G. 2006. The evolutionary psychology of facial beauty. Annu. Rev. Psychol. 57(1): 199–226.

[20] Thiruchselvam, R., J. Harper and A. L. Homer. 2016. Beauty is in the belief of the beholder: cognitive influences on the neural response to facial attractiveness. Social Cognitive and Affective Neuroscience 11(12): 1999–2008.

[21] Manning, J. T., D. Scutt and D. I. Lewis-Jones. 1998. Developmental stability, ejaculate size, and sperm quality in men. Evolution and Human Behavior 19(5): 273–282.

[22] Little, A. C., B. C. Jones, I. S. Penton-Voak, D. M. Burt and D. I. Perrett. 2002. Partnership status and the temporal context of relationships influence human female preferences for sexual dimorphism in male face shape. Proceedings of the Royal Society of London. Series B: Biological Sciences 269(1496): 1095–1100.

[23] Scott, I. M., N. Pound, I. D. Stephen, A. P. Clark and I. S. Penton-Voak. 2010. Does masculinity matter? The contribution of masculine face shape to male attractiveness in humans. PLoS One 5(10): e13585.

[24] Penton-Voak, I. S., A. C. Little, B. C. Jones, D. M. Burt, B. P. Tiddeman and D. I. Perrett. 2003. Female condition influences preferences for sexual dimorphism in faces of male humans (Homo sapiens). Journal of Comparative Psychology 117(3): 264.

[25] Cooper, J. 2001. Diagnostic and statistical manual of mental disorders (4th edn, text revision) (DSM–IV–TR) Washington, DC: American Psychiatric Association 2000. The British Journal of Psychiatry 179(1): 85–85.

[26] Little, A. C. and S. C. Roberts. 2012. Evolution, appearance, and occupational success. Evolutionary Psychology 10(5): 782–801.

[27] Lindell, A. K. and K. L. Lindell. 2014. Beauty captures the attention of the beholder. Journal of Cognitive Psychology 26(7): 768–780.

[28] Rolls, E. T. 2023. The orbitofrontal cortex, amygdala, reward value, emotion, and decision-making. Brain Computations and Connectivity. Oxford University Press. https://doi. org/10.1093/oso/9780198887911.003, 11.

[29] Chatterjee, A., A. Thomas, S. E. Smith and G. K. Aguirre. 2009. The neural response to facial attractiveness. Neuropsychology 23(2): 135.

[30] Chen, A. C., C. German and D. W. Zaidel. 1997. Brain asymmetry and facial attractiveness: Facial beauty is not simply in the eye of the beholder. Neuropsychologia 35(4): 471–476.

[31] Chuan-Peng, H., Y. Huang, S. B. Eickhoff, K. Peng and J. Sui. 2020. Seeking the "beauty center" in the brain: A meta-analysis of fMRI studies of beautiful human faces and visual art. Cognitive, Affective, and Behavioral Neuroscience 20(6): 1200–1215.

[32] Buss, D. M. and M. Barnes. 1986. Preferences in human mate selection. Journal of Personality and Social Psychology 50(3): 559.

[33] Smith, R. M. 2018. The Biology of Beauty: The Science Behind Human Attractiveness. Bloomsbury Publishing USA.

[34] Tatarunaite, E., R. Playle, K. Hood, W. Shaw and S. Richmond. 2005. Facial attractiveness: a longitudinal study. American Journal of Orthodontics and Dentofacial Orthopedics 127(6): 676–682.
[35] Ebner, N. C. 2008. Age of face matters: Age-group differences in ratings of young and old faces. Behavior Research Methods 40(1): 130–136.
[36] Berry, D. S. and L. Z. McArthur. 1985. Some components and consequences of a babyface. Journal of Personality and Social Psychology 48(2): 312.
[37] Russell, R., J. R. Sweda, A. Porcheron and E. Mauger. 2014. Sclera color changes with age and is a cue for perceiving age, health, and beauty. Psychology and Aging 29(3): 626.
[38] Tomkinson, G. R. and T. S. Olds. 2000. Physiological correlates of bilateral symmetry in humans. International Journal of Sports Medicine 21(08): 545–550.
[39] Zhao, A. A., K. Harrison, A. Holland, H. M. Wainwright, J. M. Ceccato, M. J. Sidari et al. 2023. Objectively measured facial traits predict in-person evaluations of facial attractiveness and prosociality in speed-dating partners. Evolution and Human Behavior 44(4): 315–323.
[40] Santhouse, A. M., R. J. Howard and D. H. Ffytche. 2000. Visual hallucinatory syndromes and the anatomy of the visual brain. Brain 123(10): 2055–2064.
[41] Whalen, P. J., J. Kagan, R. G. Cook, F. C. Davis, H. Kim, S. Polis et al. 2004. Human amygdala responsivity to masked fearful eye whites. Science 306(5704): 2061–2061.
[42] Hutson, J. 2020. Gallucci's Commentary on Dürer's' Four Books on Human Proportion': Renaissance Proportion Theory, 226, Open Book Publishers.
[43] Fessler, D. M., D. Nettle, Y. Afshar, I. D. A. Pinheiro, A. Bolyanatz, M. B. Mulder et al. 2005. A cross-cultural investigation of the role of foot size in physical attractiveness. Archives of Sexual Behavior 34(3): 267–276.
[44] Voracek, M., M. L. Fisher, B. Rupp, D. Lucas and D. M. Fessler. 2007. Sex differences in relative foot length and perceived attractiveness of female feet: relationships among anthropometry, physique, and preference ratings. Perceptual and Motor Skills 104(3_suppl): 1123–1138.
[45] Jones, A. L., A. Porcheron and R. Russell. 2018. Makeup changes the apparent size of facial features. Psychology of Aesthetics, Creativity, and the Arts 12(3): 359.
[46] Samson, N., B. Fink and P. J. Matts. 2010. Visible skin condition and perception of human facial appearance. International Journal of Cosmetic Science 32(3): 167–184.
[47] Kushi, M. 1983. Your Face Never Lies: What Your Face Reveals About You and Your Health, an Introduction to Oriental Diagnosis. Penguin.
[48] Russell, R., C. Batres, S. Courrèges, G. Kaminski, F. Soppelsa, F. Morizot et al. 2019. Differential effects of makeup on perceived age. British Journal of Psychology 110(1): 87–100.
[49] Fink, B., K. Grammer and P. J. Matts. 2006. Visible skin color distribution plays a role in the perception of age, attractiveness, and health in female faces. Evolution and Human Behavior 27(6): 433–442.
[50] Porcheron, A., E. Mauger and R. Russell. 2013. Aspects of facial contrast decrease with age and are cues for age perception. PloS One 8(3): e57985.
[51] Wolff, E. F., T. Altun, R. Madankumar, L. Pal, N. Santoro and H. S. Taylor. 2008. Skin wrinkles and rigidity in early postmenopausal women vary by race/ethnicity: Baseline characteristics of participants enrolled in the skin ancillary study of the keeps trial. Fertility and Sterility 90: S252.
[52] Honigman, R. J., K. A. Phillips and D. J. Castle. 2004. A review of psychosocial outcomes for patients seeking cosmetic surgery. Plastic and Reconstructive Surgery 113(4): 1229–1237.
[53] Eagly, A. H., R. D. Ashmore, M. G. Makhijani and L. C. Longo. 1991. What is beautiful is good, but…: A meta-analytic review of research on the physical attractiveness stereotype. Psychological Bulletin 110(1): 109.
[54] Dion, K., E. Berscheid and E. Walster. 1972. What is beautiful is good. Journal of Personality and Social Psychology 24(3): 285.
[55] Martín-Loeches, M., J. A. Hernández-Tamames, A. Martín and M. Urrutia. 2014. Beauty and ugliness in the bodies and faces of others: An fMRI study of person esthetic judgement. Neuroscience 277: 486–497.
[56] Tsukiura, T. and R. Cabeza. 2011. Remembering beauty: Roles of orbitofrontal and hippocampal regions in successful memory encoding of attractive faces. Neuroimage 54(1): 653–660.

[57] Aharon, I., N. Etcoff, D. Ariely, C. F. Chabris, E. O'connor and H. C. Breiter. 2001. Beautiful faces have variable reward value: fMRI and behavioral evidence. Neuron 32(3): 537–551.
[58] Berscheid, E. and H. T. Reis. 1998. Attraction and close relationships. *In*: Gilbert, D. T., S. T. Fiske and G. Lindzey (eds.). The Handbook of Social Psychology, 4: 193–281. McGraw-Hill.
[59] Berscheid, E. and E. Walster. 1974. Physical attractiveness. In Advances in Experimental Social Psychology, 7: 157–215. Academic Press.
[60] Feingold, A. 1990. Gender differences in effects of physical attractiveness on romantic attraction: A comparison across five research paradigms. Journal of Personality and Social Psychology 59(5): 981.
[61] Hosoda, M., E. F. Stone-Romero and G. Coats. 2003. The effects of physical attractiveness on job-related outcomes: A meta-analysis of experimental studies. Personnel Psychology 56(2): 431–462.
[62] Makowski, D., A. S. Te, A. Neves, S. Kirk, N. Z. Liang, P. Mavros et al. 2025. Too beautiful to be fake: Attractive faces are less likely to be judged as artificially generated. Acta Psychologica 252: 104670.
[63] Buunk, B. P., P. Dijkstra, D. Fetchenhauer and D. T. Kenrick. 2002. Age and gender differences in mate selection criteria for various involvement levels. Personal Relationships 9(3): 271–278.
[64] Fink, B., N. Neave, J. T. Manning and K. Grammer. 2006. Facial symmetry and judgements of attractiveness, health and personality. Personality and Individual Differences 41(3): 491–499.
[65] Qi, Y. and J. Ying. 2022. Gender biases in the accuracy of facial judgments: facial attractiveness and perceived socioeconomic status. Frontiers in Psychology 13: 884888.
[66] Dion, K. K. 1981. Physical attractiveness, sex roles and heterosexual attraction. The Bases of Human Sexual Attraction 3–22.
[67] Dion, K. K. 2022. Stereotyping based on physical attractiveness: Issues and conceptual perspectives. pp. 7–22. In Physical Appearance, Stigma, and Social Behavior. Routledge.
[68] Bassili, J. N. 1981. The attractiveness stereotype: Goodness or glamour? Basic and Applied Social Psychology 2(4): 235–252.
[69] Cash, T. F. and L. H. Janda. 1984. The eye of the beholder. Psychology Today 18(12): 46–52.
[70] Ashmore, R. D. 2015. Sex stereotypes and implicit personality theory. pp. 37–81. In Cognitive Processes in Stereotyping and Intergroup Behavior. Psychology Press.

Chapter 13

Brain's Perception of Gender Reassignment Surgery with Psychological Changes Based on Pre- and Post-observations

Percin Karakol,[1,*] *Mehmet Balci*,[1] *Emre Tokgoz*[2] and *Sophia Marra*[3]

13.1 Introduction

The development of the gender identity is initiated with the physiological sex development at intrautirine level; at the prenatal period, the prenatal gonads are observed to be the first sex organs to differentiate into male and female with no sexual difference that can be observed until the 6th week of the embryonic life in humans [41, 42]. There are certain clear differences between male and female brains' anomatical differences such as cortical thickness and total brain volume [43, 44]. Brain's sexual distinguishment is a complex task that involves genetic processing at a pre-hormonal level; Furthermore, development during the natural adrenarche body processing phase results in staring the production of a hormone called dehydroepiandrosterone (DHEA) at the adrenal glands that typically happens between ages 6 and 8 years of old. DHEA is known to be a precursor hormone of androgens (testosterone and androstenedione sex hormones) and estrogen that is taking place during the body development. During adrenarche processing, adrenal cortex furthermore matures in the child's adrenal glands

[1] Başakşehir Cam and Sakura City Hospital Plastic Reconstructive and Aesthetic Surgery, Istanbul, Turkiye.

[2] School of Engineering and Technology, State University of New York, Farmingdale, New York, 11735, USA.

[3] Frank H. Netter School of Medicine, Quinnipiac University, North Haven, CT, 06473, USA.

* Corresponding author: ppercin@gmail.com

DOI: 10.1201/9781003646044-13

and the cortex is the outer part of the adrenal gland that releases certain hormones; hence, all such locations play roles in child's furthermore development [45]. During development of the embryo, gender identity and the biological sex development variations can occur based on the external sex hormone exposure that are observed in the research literature [46, 47]. Among such observations, *intersex* is one of the conditions that is observed to be the case that external genitalia does not fit in the binary gender classification as male or female [48].

The brain's development with the variations in hormone-based organization influences gender identity development with the involvement of the central nervous system in such development [49]. Improper orchestration of the nervous system along with the hormonal exchange between the brain and the genitalia can cause challenges in the sex development [50].

Gender Dysphoria (GD) is a condition defined as a marked incongruence between the gender an individual experiences or expresses and the gender assigned at birth. This often leads to significant problems and dysfunctions in various areas of life. Gender reassignment surgery (GRS) is considered an important operation for individuals with severe gender dysphoria. This complex process is not only a physical transformation but also has profound effects on brain perception and psychological well-being. This review examines the multifaceted effects of CGA on brain perception, neurobiology and psychological health and aims to provide a comprehensive context of these complex processes.

Furthermore advancements and findings are developed in literature to understand the individuals' desire to make such changes that we cover in the rest of this work. The next section focuses on the neurological foundations of gender identity while the third section outlines the psychological and cognitive effects before gender change surgery. The fourth section covers postoperative neuroplastic and brain adoption perspectives. What follows in sections five and six are focusing on psychological well-being and cognitive functions, and self-perception. Section seven focuses on social and legal implications. The last section covers conclusive remarks.

13.2 Neurobiological Foundations of Gender Identity

Gender identity is a result of psychological sense of one's gender and it is a very complex systematic product of our experiences and psychological insights of who the person feels to be and defines own gender [27]. Today's variational definitions of gender identity depend on the individual's intrinsic perception of their identity among a variety of categories including the following [28, 29, 30]:

- Cisgender defined as the individual's identity selection assigned at birth.
- Transgender defined as gender selection that is different than birth intermittently or consistently

Males and females' brains have similarities, but they are dissimilar [30]. There are many studies focusing on such differences with the observations on size of specific regions or nuclei. Sex differences are observed to be impacted by synaptic pattern differentials [31, 32] along with density of neurons [33–35].

The neurobiological foundations of gender identity suggest that gender is deeply linked to the structure and function of the brain. Research shows that there are structural and functional differences in certain brain regions between individuals who experience gender dysphoria and those who do not. For example, the bed nucleus of the stria terminalis (BSTc), known to be involved in gender identity, has dimensions more aligned with the individual's identity rather than the sex assigned at birth [1, 2]. This neuroanatomical evidence suggests that gender identity is not just a psychological construct but may also have underlying contents in the structure of the brain, in which case the psychological factors that impact the formation of the brain can also be causing such formation and structures.

An accelerating number of studies in the neuroimaging literature significantly support the thesis that, again, with appropriate training and effort, people can systematically alter neural circuitry associated with a variety of mental and physical states that are pathological [36–39]; In such regard, cerebral functionality may be altered based on the "self-directed neuroplasty" that coins with one's focused effort and training to alter such cerebral functions [40].

Individuals with GD and controls are observed to have gray matter volume and brain activation differentials as well as connectivity differences as a means of a neurobiological differentiation caused by GD. The feedback that GD experiencing individuals experience conflict with their mental state of gender causes cognitive dissonance [41]. The external interactions of individuals experiencing GD also have neurobiological basis based on self-perception and the feedback received in return. A model for the development of GD including cognitive dissonance, involving anterior cingulate cortex and ventral striatum as the key brain structures is introduced for generating testable hypotheses using behavioral and neuroimaging techniques to understand the neuro-psycho-biology of GD [41].

Research has also revealed differences in the white matter microstructure of transgender individuals compared to cisgender individuals. These differences were observed in brain regions related to body perception and self-awareness, suggesting that transgender individuals have a neurobiological bias toward the gender they experience [3, 4]. Such research was critical towards understanding the neurobiological underpinnings of gender identity.

13.3 Psychological and Cognitive Effects Before Surgery

Prior to gender transition surgery, individuals with gender dysphoria often face significant psychological problems. These problems stem from a mismatch between gender identity and physical appearance, leading to anxiety, depression and a persistent cognitive dissonance [5]. The brain is in a constant effort to reconcile this dissonance, resulting in altered patterns in neural activity.

These patterns often manifest as increased activity in brain areas involved in self-identity thinking and body image processing [6, 7].

Research has shown that transgender individuals process gender-specific stimuli differently from cisgender individuals. For example, studies using functional MRI have found that transgender women (assigned male at birth) show more similar brain activation patterns to stimuli such as gendered faces and voices compared to cisgender women [8]. This suggests that the way the brain perceives gender-specific stimuli is congruent with an individual's gender identity, even before surgery.

The psychological pressure of living with gender dysphoria is high. Many transgender individuals experience increased anxiety, depression and suicidal thoughts, and decreased quality of life [9]. These persistent cognitive problems can also impair functions such as memory and focus [10]. These psychological and cognitive difficulties emphasize the need for interventions that aim to harmonize physical appearance with gender identity.

13.4 Postoperative Neuroplasticity and Brain Adaptation

Gender reassignment surgery (GRS) related long-term studies are limited with the typical observation period to be less than 1 year [11] on limited number of individuals. The results reported on longer term reviews (i.e., more than 1-year) on the post-operative pain experience are sparse with conflicting results [12]. While some improvements have been reported by some of the individuals who experienced GRS in the research literature based on post-operative GRS experiences [11], the contrary was also observed in some other studies.

The studies that lasted longer than a year-long in determining post-operative pain effects of GRS is very limited in their scope; Studies that conducted follow up on completed gender affirming surgeries indicated high suicidal rates and psychiatric challenges while gender dysphoria's somewhat improvement [15, 16]. Research shows that post-surgery transgender individuals' brain responses to gender-specific stimuli causes changes particularly in regions such as the insula and anterior cingulate cortex, which are involved in processing self-referential and emotional information [13]. Hormonal treatment also plays a critical role in brain adaptation. Testosterone and estrogen have been shown to cause changes in brain structure, regions such as the hypothalamus and white matter tracts [14].

13.5 Psychological Well-Being Considerations

There may be psychological challenges and benefits faced as a result of gender transition surgery after its completion; There are studies that reported improvements in body satisfaction, social functioning and overall quality of life after surgery [17, 18] as well as the alignment of physical appearance with gender identity is mentioned to cause a significant reduction in anxiety and depression and improvements in self-esteem and self-concept [19]; However, on the flip side, it can be causing family, friend, and peer relationship conflicts that may lead to rejection from society along with symptoms of depression and anxiety, poor self-esteem, substance use disorder increased risk of self-harm and suicidality, and interpersonal conflicts [53].

On the improvement side, for psychological aspects is the normalization of brain responses to self-related stimuli. Post-surgery, transgender people often feel more comfortable with their bodies and experience a sense of congruence between physical appearance and gender identity [20]. This sense of congruence is important for the development of stable and consistent self-perception, which is critical for long-term psychological well-being. On the flip side, the statistics based on the collected survey data in the United States indicate increasing challenges faced in a different way.

Results reported in 2017 as a part of a sexual health study revealed an increase in the self-reported transgender identity in adults, adolescents and children anywhere between 0.5–1.3% and the increase was identified to be occurring within the last few decades [54]. Approximately 4 in 1,000 adults were determined to be transgender in a 2016 national survey data collected based on the data attained from 100,000 adults with a remark that the number may be an underestimation of the reality [55]. Approximately 1.4 million Americans that represented approximately 0.6% of the U.S.A. population identified as transgender while European estimation ranged from 0.8–1.1% for assigned females [59–61]. Such an increase in the Unites States has been in correlation with the population's seek of health over the last decade [56].

Research results are reported on certain challenges faced by the individuals that experienced gender dysphoria [57, 58]:

- Substance use disorder on 38% of observed study participants.
- Suicidal ideation of 48.3% study participants with 23.8% of them already attempted at least one time suicide.
- Personality disorder was observed for 52% of the participants in a study with cluster B personality disorder to be the most prevalent one.

Interactions with family during childhood based on children's temperament and parental qualities with their impact on the dynamic relationship established can be significant influencers of a child's that starts gender identification typically between 3–5 years old. There is extensive relevant study and research conducted in relevant areas [53, 62].

13.6 Cognitive Functions and Self-Perception

Post-surgery, many transgender individuals experience improvements in their cognitive functioning, particularly in areas such as memory, attention and executive functioning [21]. The reduction of cognitive dissonance and alleviation of psychological distress likely contribute to these improvements. Because the brain no longer has to constantly reconcile the incongruence between physical appearance and gender identity, cognitive resources can be redirected to more adaptive functions [22].

Self-perception also undergoes significant change after surgery. The brain's internal body representation, the "body schema", is reshaped to adapt to the new physical reality [23]. This adaptation process is critical for a sense of body ownership and wholeness [24]. Over time, this process leads to a more stable and coherent self-perception, which is critical for long-term psychological well-being.

13.7 Social and Legal Implications

The social and legal dimensions of gender transition are also critical in shaping the psychological outcomes of transgender individuals. Transgender people face many challenges, including access to healthcare, legal recognition of gender identity, and protection from discrimination [25, 26]. These challenges can have a significant impact on mental health, so social and legal barriers need to be considered to realize the full benefits of gender transition surgery.

One such legal implication is related to identity change relates to the same sex marriage banning depending on the country and even the specific state in the United States. For instance, while Netherland allows same sex marriage, many other countries prohibit the marriage of people of same sex [51].

Another legal implication is related to the medical professional's sex assignment as legal institutions require humans to have a binary categorization; The relevant implications include but not limited to the following [52]:

- Gender classification on legal documents
- Marriage
- Athletic competition gender classification
- Discrimination related considerations
- Sex-based crimes
- Prisoning based on gender
- Constitutional equality related rights
- Payments required to be made based on gender classification (such as insurance payments)

It is also essential to note that there are many more social and legal implications that include but not limited to the coverage in this work therefore we refer to [29, 47, 51, 52] for furthermore reading on the corresponding contents.

13.8 Conclusion

Gender transition surgery represents a profound transformation, encompassing both physical and psychological dimensions. Changes in brain perception before and after surgery highlight the deep connection between gender identity, brain structure and cognitive processes. An area of research that may be difficult to conduct and measure is identification of the correlation between psychological problems faced along with the physiological implications based on the underlying circumstances that the individual had in life. The complexity of the life conditions and challenges experienced by a person who decides to undergo a gender transition surgery is a tough area of research as such research cannot observe similar conditions with similarities in life background, psychological, and physiological factors. Before surgery, the person's thinking process is often in a state of great conflict as the individual is trying to reconcile the mismatch between physical appearance and gender identity. After surgery, however, the psychological conditions can shift to a brain adaptation to the new physical reality that leads to a new psychological reality with the corresponding changes in cognitive functioning. In this case, the research question may need to target investigating whether it is the psychological relief the individual seeking or the physiological changes that were needed. This is a complex phenomenon that is not easy to investigate. In alignment with this, one may not necessarily know the traumatic events that may have been experienced by an individual who desires to have gender transition surgery that may lead to such a decision and possible physiological changes in the brain.

There is a good number of neuroimaging studies significantly supporting how appropriate training and efforts can systematically alter neural circuitry associated with a variety of mental and physical states. The studies on long term effects of surgery are not comprehensively covering the necessary details, and such a study would be value-adding to furthermore understand the long-term aftermath of such studies. Understanding post-surgery changes is critical to providing comprehensive care for transgender people. As research in this area develops, it will offer deeper insights into the relationship between gender identity, brain structure and psychological processes, which will contribute to better support and treatment options for individuals undergoing gender transition surgery.

References

[1] Zhou, J. N., M. A. Hofman, L. J. Gooren and D. F. Swaab. 1995. A sex difference in the human brain and its relation to transsexuality. Nature 378(6552): 68–70.

[2] Kruijver, F. P., J. N. Zhou, C. W. Pool, M. A. Hofman, L. J. Gooren and D. F. Swaab. 2000. Male-to-female transsexuals have female neuron numbers in a limbic nucleus. The Journal of Clinical Endocrinology and Metabolism 85(5): 2034–2041.

[3] Rametti, G., B. Carrillo, E. Gómez-Gil, C. Junque, L. Zubiaurre-Elorza, S. Segovia et al. 2012. Effects of androgenization on the white matter microstructure of female-to-male transsexuals. A diffusion tensor imaging study. Psychoneuroendocrinology 37(8): 1261–1269.

[4] Guillamon, A., C. Junque and E. Gómez-Gil. 2016. A review of the status of brain structure research in transsexualism. Archives of Sexual Behavior 45(7): 1615–1648.

[5] DSM-5, American Psychiatric Association. 2013. https://psychiatryonline.org/doi/book/10.1176/appi.books.9780890425596.

[6] Spack, N. P., L. Edwards-Leeper, H. A. Feldman, S. Leibowitz, F. Mandel, D. A. Diamond et al. 2012. Children and adolescents with gender identity disorder referred to a pediatric medical center. Pediatrics 129(3): 418–425.

[7] Smith, Y. L., S. H. Van Goozen, A. J. Kuiper and Cohen-Kettenis. 2005. Sex reassignment: outcomes and predictors of treatment for adolescent and adult transsexuals. Psychological Medicine 35(1): 89–99.

[8] Savic, I. and S. Arver. 2011. Sex dimorphism of the brain in male-to-female transsexuals. Cerebral Cortex 21(11): 2525–2533.

[9] Dhejne, C., P. Lichtenstein, M. Boman, A. L. Johansson, N. Långström and M. Landén. 2011. Long-term follow-up of transsexual persons undergoing sex reassignment surgery: cohort study in Sweden. PloS One 6(2): e16885.

[10] Murad, M. H., M. B. Elamin, M. Z. Garcia, R. J. Mullan, A. Murad, P. J. Erwin et al. 2010. Hormonal therapy and sex reassignment: A systematic review and meta-analysis of quality of life and psychosocial outcomes. Clinical Endocrinology 72(2): 214–231.

[11] Anger, J. T., L. K. Case, A. P. Baranowski, A. Berger, R. M. Craft, L. A. Damitz et al. 2024. Pain mechanisms in the transgender individual: a review. Frontiers in Pain Research 5: 1241015.

[12] Lindqvist, E. K., H. Sigurjonsson, C. Möllermark, J. Rinder, F. Farnebo and T. K. Lundgren. 2017. Quality of life improves early after gender reassignment surgery in transgender women. European Journal of Plastic Surgery 40(3): 223–226.

[13] Pol, H. E. H., P. T. Cohen-Kettenis, N. E. Van Haren, J. S. Peper, R. G. Brans, W. Cahn et al. 2006. Changing your sex changes your brain: influences of testosterone and estrogen on adult human brain structure. European Journal of Endocrinology 155(Supplement_1): S107–S114.

[14] Bentz, E. K., L. A. Hefler, U. Kaufmann, J. C. Huber, A. Kolbus and C. Tempfer. 2008. A polymorphism of the CYP17 gene related to sex steroid metabolism is associated with female-to-male but not male-to-female transsexualism. Fertility and Sterility 90(1): 56–59.

[15] Park, R. H., Y. T. Liu, A. Samuel, M. Gurganus, T. J. Gampper and S. T. Corbett. 2022. Long-term outcomes after gender-affirming surgery: 40-year follow-up study. Annals of Plastic Surgery 89(4): 431–436.

[16] Dhejne, C., P. Lichtenstein, M. Boman, A. L. Johansson, N. Långström and M. Landén. 2011. Long-term follow-up of transsexual persons undergoing sex reassignment surgery: cohort study in Sweden. PloS One 6(2): e16885.

[17] Zubiaurre-Elorza, L., C. Junque, E. Gómez-Gil, S. Segovia, B. Carrillo, G. Rametti et al. 2013. Cortical thickness in untreated transsexuals. Cerebral Cortex 23(12): 2855–2862.

[18] Gijs, L. and A. Brewaeys. 2007. Surgical treatment of gender dysphoria in adults and adolescents: Recent developments, effectiveness, and challenges. Annual Review of Sex Research 18(1): 178–224.

[19] Dhejne, C., K. Öberg, S. Arver and M. Landén. 2014. An analysis of all applications for sex reassignment surgery in Sweden, 1960–2010: Prevalence, incidence, and regrets. Archives of Sexual Behavior 43(8): 1535–1545.

[20] Haraldsen, I. R. and A. A. Dahl. 2000. Symptom profiles of gender dysphoric patients of transsexual type compared to patients with personality disorders and healthy adults. Acta Psychiatrica Scandinavica 102(4): 276–281.

[21] Guillamon, A., C. Junque and E. Gómez-Gil. 2016. A review of the status of brain structure research in transsexualism. Archives of Sexual Behavior 45(7): 1615–1648.

[22] Kreukels, B. P. and P. T. Cohen-Kettenis. 2011. Puberty suppression in gender identity disorder: the Amsterdam experience. Nature Reviews Endocrinology 7(8): 466–472.

[23] Swaab, D. F. 2007. Sexual differentiation of the brain and behavior. Best Practice and Research Clinical Endocrinology and Metabolism 21(3): 431–444.

[24] Blanchard, R., L. H. Clemmensen and B. W. Steiner. 1987. Heterosexual and homosexual gender dysphoria. Archives of Sexual Behavior 16(2): 139–152.

[25] Bockting, W., A. Benner and E. Coleman. 2009. Gay and bisexual identity development among female-to-male transsexuals in North America: Emergence of a transgender sexuality. Archives of Sexual Behavior 38(5): 688–701.

[26] Spade, D. 2015. Normal Life: Administrative Violence, Critical Trans Politics, and The Limits of Law. Duke University Press.

[27] Ristori, J., C. Cocchetti, A. Romani, F. Mazzoli, L. Vignozzi, M. Maggi et al. 2020. Brain sex differences related to gender identity development: genes or hormones?. International Journal of Molecular Sciences 21(6): 2123.

[28] Polderman, T. J., B. P. Kreukels, M. S. Irwig, L. Beach, Y. M. Chan, E. Derks et al. 2018. International Gender Diversity Genomics Consortium. The biological contributions to gender identity and gender diversity: bringing data to the table. Behavior Genetics 48(2): 95–108.

[29] Brumbaugh-Johnson, S. M. and K. E. Hull. 2019. Coming out as transgender: Navigating the social implications of a transgender identity. Journal of Homosexuality 66(8): 1148–1177.

[30] Ngun, T. C., N. Ghahramani, F. J. Sánchez, S. Bocklandt and E. Vilain. 2011. The genetics of sex differences in brain and behavior. Frontiers in Neuroendocrinology 32(2): 227–246.

[31] Greenough, W. T., C. S. Carter, C. Steerman and T. J. DeVoogd. 1977. Sex differences in dendritic patterns in hamster preoptic area. Brain Research 126(1): 63–72.
[32] Ciofi, P., D. Leroy and G. Tramu. 2006. Sexual dimorphism in the organization of the rat hypothalamic infundibular area. Neuroscience 141(4): 1731–1745.
[33] Good, C. D., I. Johnsrude, J. Ashburner, R. N. Henson, K. J. Friston and R. S. Frackowiak. 2001. Cerebral asymmetry and the effects of sex and handedness on brain structure: a voxel-based morphometric analysis of 465 normal adult human brains. Neuroimage 14(3): 685–700.
[34] Matsumoto, A. and Y. Arai. 1986. Male-female difference in synaptic organization of the ventromedial nucleus of the hypothalamus in the rat. Neuroendocrinology 42(3): 232–236.
[35] Witelson, S. F., I. I. Glezer and D. L. Kigar. 1995. Women have greater density of neurons in posterior temporal cortex. Journal of Neuroscience 15(5): 3418–3428.
[36] Schwartz, J. M., P. W. Stoessel, L. R. Baxter, K. M. Martin and M. E. Phelps. 1996. Systematic changes in cerebral glucose metabolic rate after successful behavior modification treatment of obsessive-compulsive disorder. Archives of General Psychiatry 53(2): 109–113.
[37] Schwartz, J. M. 1998. Neuroanatomical aspects of cognitive-behavioural therapy response in obsessive-compulsive disorder: An evolving perspective on brain and behaviour. The British Journal of Psychiatry 173(S35): 38–44.
[38] Musso, M., C. Weiller, S. Kiebel, S. P. Müller, P. Bülau and M. Rijntjes. 1999. Training-induced brain plasticity in aphasia. Brain 122(9): 1781–1790.
[39] Paquette, V., J. Lévesque, B. Mensour, J. M. Leroux, G. Beaudoin, P. Bourgouin et al. 2003. "Change the mind and you change the brain": effects of cognitive-behavioral therapy on the neural correlates of spider phobia. Neuroimage 18(2): 401–409.
[40] Schwartz, J. and S. Begley. 2002. The Mind and the Brain: Neuroplasticity and the Power of Mental Force p. 118. New York: Regent.
[41] Altinay, M. and A. Anand. 2020. Neuroimaging gender dysphoria: a novel psychobiological model. Brain Imaging and Behavior 14(4): 1281–1297.
[42] Ray, R. and C. Racine. 2025. Sexual Differentiation. [Updated 2025 Jul 24]. *In*: Feingold, K. R., S. F. Ahmed, B. Anawalt et al. (eds.). Endotext [Internet]. South Dartmouth (MA): MDText.com, Inc.; 2000-. Available from: https://www.ncbi.nlm.nih.gov/books/NBK279001/.
[43] Swaab, D. F. and M. A. Hofman. 1984. Sexual differentiation of the human brain a historical perspective. Progress in Brain Research 61: 361–374.
[44] Bramen, J. E., J. A. Hranilovich, R. E. Dahl, J. Chen, C. Rosso, E. E. Forbes et al. 2012. Sex matters during adolescence: testosterone-related cortical thickness maturation differs between boys and girls. PloS One 7(3): e33850.
[45] Klinge, C. M., B. J. Clark and R. A. Prough. 2018. Dehydroepiandrosterone research: past, current, and future. Vitamins and Hormones 108: 1–28.
[46] Kerlin, S. 2005. Prenatal exposure to diethylstilbestrol (DES) in males and gender-related disorders: results from a 5-year study. In Int Behav Dev Symp (Vol. 1).
[47] Bailey, J. M., M. P. Dunne and N. G. Martin. 2000. Genetic and environmental influences on sexual orientation and its correlates in an Australian twin sample. Journal of Personality and Social Psychology 78(3): 524.
[48] Hughes, I. A., C. Houk, S. F. Ahmed, P. A. Lee and L. W. P. E. Society. 2006. Consensus statement on management of intersex disorders. Journal of Pediatric Urology 2(3): 148–162.
[49] Schulz, K. M., H. A. Molenda-Figueira and C. L. Sisk. 2009. Back to the future: the organizational–activational hypothesis adapted to puberty and adolescence. Hormones and Behavior 55(5): 597–604.
[50] Roselli, C. E. 2018. Neurobiology of gender identity and sexual orientation. Journal of Neuroendocrinology 30(7): e12562.
[51] Greenberg, J. A. 1999. Defining male and female: Intersexuality and the collision between law and biology. Ariz. L. Rev. 41: 265.
[52] Greenberg, J. 2003. Legal aspects of gender assignment. The Endocrinologist 13(3): 277–286.
[53] Garg, G., G. Elshimy and R. Marwaha. 2023. Gender Dysphoria. [Updated 2023 Jul 11]. In: StatPearls [Internet]. Treasure Island (FL): StatPearls Publishing; 2025 Jan-. Available from: https://www.ncbi.nlm.nih.gov/books/NBK532313/.

[54] Zucker, K. J. 2017. Epidemiology of gender dysphoria and transgender identity. Sexual Health 14(5): 404–411.
[55] Meerwijk, E. L. and J. M. Sevelius. 2017. Transgender population size in the United States: a meta-regression of population-based probability samples. American Journal of Public Health 107(2): e1–e8.
[56] Polderman, T. J., B. P. Kreukels, M. S. Irwig, L. Beach, Y. M. Chan, E. M. Derks et al. 2018. International Gender Diversity Genomics Consortium. The biological contributions to gender identity and gender diversity: bringing data to the table. Behavior Genetics 48(2): 95–108.
[57] García-Vega, E., A. Camero, M. Fernández and A. Villaverde. 2018. Suicidal ideation and suicide attempts in persons with gender dysphoria. Psicothema 30(3): 283–288.
[58] Madeddu, F., A. Prunas and D. Hartmann. 2009. Prevalence of Axis II disorders in a sample of clients undertaking psychiatric evaluation for sex reassignment surgery. Psychiatric Quarterly 80(4): 261–267.
[59] Bakker, A., P. J. van Kesteren, L. J. Gooren and P. D. Bezemer. 1993. The prevalence of transsexualism in the Netherlands. Acta Psychiatrica Scandinavica 87(4): 237–238.
[60] Kuyper, L. and C. Wijsen. 2014. Gender identities and gender dysphoria in the Netherlands. Archives of Sexual Behavior 43(2): 377–385.
[61] Reed, B., S. Rhodes, P. Schofield and K. Wylie. 2009. Gender variance in the UK: Prevalence, incidence, growth and geographic distribution.
[62] Heller, S. 2024. A Psychoanalytic Exploration of Female to Male Transition. Taylor and Francis.

Chapter 14

Adaptation of the Brain After Limb Amputations

Percin Karakol,[1,*] *Huseyin Ulas,*[1] *Emre Tokgoz*[2] and *Sophia Marra*[3]

14.1 Introduction

Neuronal changes, alterations in neurotransmitter levels, and physiological as well as psychiatric factors play a significant role in the brain's adaptation to limb amputations [1, 2]. During the post-amputation period, a multidisciplinary approach is essential, encompassing physical rehabilitation, psychological support, and cognitive training [2]. The involvement of surgeons, physical medicine and rehabilitation specialists, therapists, prosthetists, and pain management teams is crucial in providing comprehensive care. This holistic approach is vital in ensuring the best possible support and treatment for patients, addressing both their physical and psychological needs [1, 2, 3]. This multidisciplinary team is integral to facilitating the patient's adjustment process, enhancing functional outcomes, and improving overall well-being [3]. The coordinated efforts of these healthcare professionals help in managing the complexities associated with limb loss, thereby promoting recovery and adaptation [1, 2, 3].

Phantom limb is a phenomenon in which individuals who have undergone an amputation experience sensation as if the amputated limb is still present, often accompanied by vivid sensory and motor characteristics. This condition has garnered significant interest from both researchers and clinicians [4]. The modern understanding of phantom limb attributes its occurrence to physiological changes

[1] Başakşehir Cam and Sakura City Hospital Plastic Reconstructive and Aesthetic Surgery, Istanbul, Turkiye.
[2] School of Engineering and Technology, State University of New York, Farmingdale, New York, 11735, USA.
[3] Frank H. Netter School of Medicine, Quinnipiac University, North Haven, CT, 06473, USA.
* Corresponding author: ppercin@gmail.com

DOI: 10.1201/9781003646044-14

that occur following limb amputation [4]. The neurophysiological basis of the phantom limb phenomenon is primarily attributed to the activity of neurons in the brain and central nervous system, particularly involving the reorganization of the somatosensory and motor cortices [5]. The neuromatrix theory, proposed by Melzack in the early 1990s, provided a fundamental framework for explaining the brain activity underlying this phenomenon [6].

Research conducted over the last 40–50 years has demonstrated that sensory maps in the brain can indeed change following amputations, reigniting interest in the clinical phenomenon of phantom limbs. These studies have shown how new neural connections form and how information from various sensory modalities, such as touch, proprioception, and vision, interact [7]. This neuroplasticity suggests that the brain is capable of significant reorganization, adapting to changes in sensory input even in adulthood [7, 8]. This understanding has profound implications for the treatment of phantom limb pain and the development of therapeutic strategies that can harness this plasticity to improve patient outcomes [8].

Brain plasticity, defined as the brain's ability to change its structure, is particularly important in the context of somatosensory and motor reorganization following limb amputation [9, 25]. Various types of sensorimotor plasticity that emerge after amputation can involve the reorganization of nociceptive pathways associated with different components of phantom limb pain [9]. The complexity of these changes is emphasized by the interaction of multiple behavioral and neural inputs, including proprioceptive and visual feedback, tactile sensations, and cognitive factors such as reward and motivation [10]. These factors contribute to the dynamic and adaptive processes within the brain that facilitate recovery and adaptation post-amputation [10].

Functional neuroimaging techniques such as functional magnetic resonance imaging (fMRI), positron emission tomography (PET), and magnetoencephalography (MEG) have provided valuable insights into cortical reorganization associated with phantom limb pain [11, 12, 13]. These imaging methods have revealed that adjacent representation areas in the primary somatosensory cortex shift into regions that previously represented the amputated limb, a phenomenon referred to as cortical reorganization; This neural reorganization is thought to contribute to the experience of phantom limb pain [12, 13]. The changes in the brain's somatosensory map suggest that the sensory cortex remains plastic and capable of reorganization even in adulthood, which has significant implications for understanding and treating phantom limb pain [13].

In addition to neurophysiological changes, the psychosocial impacts of limb amputation are highly significant, affecting patients' mental health, social interactions, and overall quality of life [14]. They often experience high levels of clinical depression, post-traumatic stress disorder, anxiety, and long-term psychological distress [15].

14.2 Phantom Limbs

Phantom limb describes the illusion that extremity amputees believe that their amputated limbs are still present and that they have living sensory and motor qualities [16]. This phenomenon was first discovered and described by the French military surgeon Ambroise Paré, but nearly 200 years later Silas Weir Mitchell called it the "phantom limb" (1871) [16]. Mitchell believed the phantom limb represented a sensory illusion [16]. The modern understanding of the phantom limb is that this condition results from physiological changes that occur following limb amputation [17]. The physiological origin of the phantom limb arises from activity within the brain and central nervous system neurons [17]. In the early 1990s, Melzack put forward the neuromatrix theory, and the first step was taken to explain the brain activity underlying the phantom limb phenomenon [18]. He proposed that there is a dispersed network of neurons in the brain and a characteristic pattern of activity that indicates the holistic sound state of the body [18]. This neuronal network has three main neural circuit activities: a pathway from the reticular formation to the limbic system, a sensory pathway from the thalamus to the somatosensory cortex, and another pathway through the parietal lobe [18, 19]. Even if the network of neurons in the brain receives information indicating that a body part no longer exists, it cannot be updated to reflect this changing state of the body [19]. The conflict between input from the environment, which indicates that the body is no longer completely intact, and the activity of the hard-wired neuron network, which indicates that the body is whole, gives rise to the feeling that peripheral activity has undergone an error [19]. This means that the brain senses that the missing limb is still present, even though information from peripheral activity indicates otherwise [19].

The neuromatrix theory of brain function, proposed based on the phantom limb phenomenon, attempts to explain how phantom limb pain occurs [19, 20]. The neuromatrix produces a pattern of neurological signatures in the absence of modulatory inputs from the limbs [20]. In the absence of a limb, messages to move the muscles are tried to be transmitted more frequently and 'stronger' when trying to move the limb, and the final result of this message can be felt as muscle cramp pain. It turns out that the origin of phantom limb pain lies in the brain [20].

14.3 Plasticity and Reorganization of the Somatosensory and Somatomotor System

The adult primate brain has different topographically organized maps related to different forms of sensation [21]. Until very recently, it was believed that new neural connections could not form in the adult mammalian brain [22]. Various studies over the last 40–50 years on the effects of amputations on somatosensory maps have allowed us to revise this understanding of the nervous system in adult humans and other mammals [22]. These studies have shown that sensory maps can indeed change in the adult brain, sparking a revival of interest in the clinical phenomenon of phantom limbs [22]. However, they have shown not only how new connections emerge in the adult human brain, but also how information from different sensory nodules (e.g., touch, proprioception, and vision) interact [22, 23]. Studies on phantom limbs show how exactly the brain forms a body image and how this image is reflected in changing sensory receptors [23]. It also allowed us to understand how it is constantly updated in response to inputs [23].

Gross somatotopic organization is the task of the primary somatosensory and motor cortex. The cerebral cortex is systematically organized to control the body's movements [24]. Forming neural connections in response to external stimuli and various responses to them have been reported in sensorimotor areas [26]. It has been demonstrated that brain plasticity also occurs in different diseases, for example, in sensory or motor training, in immobilized patients, in patients suffering from stroke or limb amputation [24, 25, 26]. There are many types of sensorimotor plasticity that occur after the loss of a limb [27]. Although the perceptual sensations emanating from the phantom limb are still not fully understood, sensorimotor plasticity after amputation may be related to the reorganization of nociceptive pathways that are spatially and temporally associated with different components of phantom limb pain [27]. Individuals experiencing phantom limb pain have shown reduced activity in the secondary somatosensory cortex, insula, orbitofrontal cortex, anterior cingulate cortex, and striatum [28]. This complexity is also evidenced by versatile strategies that combine multiple behavioral and neural sources, such as proprioceptive and visual feedback, tactile sensations, and even more cognitive aspects such as reward and motivation [27, 28]. Research indicates that stimulating the motor cortex can potentially modify the pain experienced due to deafferentation. There is evidence linking phantom limb pain to the reorganization of somatosensory and somatomotor areas following amputation [29]. Additionally, stimulation of the somatosensory cortex has been found to enhance the motor cortex over the long term [30]. There are connection projections from the somatosensory cortex to the motor cortex, and these projections play an important role in the acquisition of motor skills [29, 30]. It follows that reorganization in the motor cortex may be secondary to changes in the somatosensory cortex [29, 30]. Another possibility is that changes in the thalamus could influence motor plasticity [30]. Following limb amputation, major reorganization occurs in the primary sensory cortex and motor cortex, and this is thought to cause phantom limb pain, poorer voluntary control transmission and higher levels of pain transmission in the amputated area due to the inability to use that limb anymore [31]. The ability to move the phantom in a group of patients with hand amputation treated with mirror or imagery therapies has been shown that synergistic movement of intact and phantom limbs can help shape and reactivate the deprived cortex. These therapies enabled voluntary movements of the phantom hand and mitigated the reorganization of the motor cortex opposite the amputation, consequently alleviating phantom limb pain [31]. Hyperexcitability observed in the cortex after unilateral amputation can result from an increase in the excitatory neurotransmitter glutamate and its AMPA and NMDA receptors, along with a decrease in GABA receptors and neurotransmitters [32]. In this case, excitation occurs more than inhibition [32]. Therefore, treatments aim to correct the excitatory/inhibitory imbalance in the brain, targeting both pain reduction and enhanced motor recovery following amputation/denervation [33].

14.4 Methods of Examining Brain Functions

Functional magnetic resonance imaging (fMRI) is an imaging method based on indirect measurement of changes in deoxyhemoglobin concentration resulting from changes in blood oxygenation, widely used in cognitive neuroscience and clinic to create images and maps of the brain's activity [34]. In a study, functional MRI in patients with left hand amputation compared the somatosensory cortex and somatomotor cortex with healthy controls during the performance of intact hand movements and phantom limb movements, and it was observed that the movement of the healthy hand in amputees showed activation at a level similar to the movement of the right dominant hand in healthy controls, but for the phantom hand, all patients exhibited significantly greater activation in the primary motor and somatosensory cortices on the side opposite the amputation compared to the activation observed when control subjects imagined hand movements [35]. Additionally, in patients with phantom limb pain, imaginary movement of the phantom hand was also observed to activate the adjacent facial region, revealing selective co-activation of cortical hand and mouth regions in patients with phantom limb pain [35]. Using PET imaging methods as well as functional MRI, it has also been shown that neighboring representation regions in the primary somatosensory cortex shift to the area that previously represented the amputated limb, and this is defined as cortical restructuring [36, 37]. This reorganization can be explained as the neural correlation of phantom limb pain [36, 37]. Magnetoencephalography (MEG) is a non-invasive imaging method developed to examine brain function, particularly underlying oscillations in neuronal activity in vivo, and involves hundreds of distortion-free sensors surrounding the head to provide high spatial sensitivity over superficial cortical areas [38]. Functional mapping of the somatosensory cortex can be achieved by functional MR imaging and/or magnetoencephalography [39]. A combination of magnetoencephalography and functional MR imaging may allow increased sensitivity, fewer false-positive results, and high spatial and temporal resolution [39].

14.5 Assessment of Cognitive Functioning and Posttraumatic Stress Disorders in Patients with Limb Amputations

The amputation of a limb has not only physical effects but also psychological and social impacts on an individual [40]. Limb loss can negatively impact patients' social lives, occupational status, daily activities, and overall life satisfaction [41]. Cognitive impairment is relatively common among individuals with an amputated limb and significantly affects functional outcomes during patient rehabilitation [42]. It is important to assess the cognitive abilities of patients in the early stages of rehabilitation to set appropriate and realistic goals and to tailor the rehabilitation program to the patients' strengths, ensuring maximum efficiency [43]. Studies have shown that a significant proportion of individuals with upper extremity amputation experience clinical depression (20–55%), post-traumatic stress disorder (23–24%), long-term post-traumatic psychological distress (25%) and anxiety (36%) after amputation [44]. In individuals with upper extremity amputation, levels of depression and anxiety have been found to be higher and psychosocial adjustment poorer compared to those with lower extremity amputation [45]. This difference is thought to be due to the more prominent visibility of the upper extremity and the difficulty of concealing the prosthesis under clothing [46]. Patients' adjustment to amputation shapes their coping strategies, social discomfort, perceived social stigma, and self-consciousness in public [47].

14.6 Conclusions

Various neuronal changes, alterations in neurotransmitter levels, and physiological and psychiatric factors play a role in the brain's adaptation to limb amputations. The phenomenon of phantom limb, experienced by amputees, encompasses the illusion of the amputated limb's continued presence and function. In modern science, it is understood that phantom limbs result from physiological changes following amputation. The neurophysiological basis of phantom limbs is attributed to brain and central nervous system activity, particularly the reorganization of the somatosensory and motor cortices. Studies using neuroimaging techniques such as functional MRI, PET, and magnetoencephalography have demonstrated that the cortical regions representing the missing limb undergo significant restructuring, which may contribute to the experience of phantom limb pain. Phantom limb pain is typically associated with hyperactivity in the cortex, which relates to increased levels of excitatory neurotransmitters (e.g., glutamate) and decreased levels of inhibitory neurotransmitters (e.g., GABA). This imbalance suggests that treatment approaches should aim to restore the excitation/inhibition balance to alleviate pain and enhance motor recovery.

The psychosocial impacts of limb amputation are profound, affecting patients' mental health, social interactions, and overall quality of life. In light of these findings, the early and comprehensive assessment of cognitive and psychosocial factors in amputee patients is of critical importance. Tailoring rehabilitation programs to individual strengths and setting realistic goals can improve functional outcomes and enhance overall well-being. The importance of a multidisciplinary approach, including psychological support and cognitive training, cannot be overstated in addressing the complex needs of amputee patients. After amputation, presence of a multidisciplinary treatment approach involving surgeons, physical medicine and rehabilitation specialists, occupational therapists, prosthetists, and pain management specialists has been shown to be crucial.

References

[1] Wheaton, L. A. 2017. Neurorehabilitation in upper limb amputation: understanding how neurophysiological changes can affect functional rehabilitation. Journal of Neuroengineering and Rehabilitation 14(1): 41. https://doi.org/10.1186/s12984-017-0256-8.

[2] Penn-Barwell, J. G. 2011. Outcomes in lower limb amputation following trauma: a systematic review and meta-analysis. Injury 42(12): 1474–1479. https://doi.org/10.1016/j.injury.2011.07.005.

[3] Musuuza, J., B. L. Sutherland, S. Kurter, P. Balasubramanian, C. M. Bartels and M. B. Brennan. 2020. A systematic review of multidisciplinary teams to reduce major amputations for patients with diabetic foot ulcers. Journal of Vascular Surgery 71(4): 1433–1446.e3. https://doi.org/10.1016/j.jvs.2019.08.244.

[4] Collins, K. L., H. G. Russell, P. J. Schumacher, K. E. Robinson-Freeman, E. C. O'Conor et al. 2018. A review of current theories and treatments for phantom limb pain. The Journal of Clinical Investigation 128(6): 2168–2176. https://doi.org/10.1172/JCI94003.

[5] Makin, T. R. and H. Flor. 2020. Brain (re)organisation following amputation: Implications for phantom limb pain. NeuroImage 218: 116943. https://doi.org/10.1016/j.neuroimage.2020.116943.

[6] Melzack, R. 2005. Evolution of the neuromatrix theory of pain. The Prithvi Raj Lecture: presented at the third World Congress of World Institute of Pain, Barcelona 2004. Pain Practice : The Official Journal of World Institute of Pain 5(2): 85–94. https://doi.org/10.1111/j.1533-2500.2005.05203.x.

[7] Grüsser, S. M., M. Diers and H. Flor. 2003. Phantomschmerz: Aspekte der Neuroplastizität und Intervention [Phantom limb pain: aspects of neuroplasticity and intervention]. Anasthesiologie, Intensivmedizin, Notfallmedizin, Schmerztherapie: AINS 38(12): 762–766. https://doi.org/10.1055/s-2003-45403.

[8] Andoh, J., C. Milde, J. W. Tsao and H. Flor. 2018. Cortical plasticity as a basis of phantom limb pain: Fact or fiction?. Neuroscience 387: 85–91. https://doi.org/10.1016/j.neuroscience.2017.11.015.

[9] Yanagisawa, T., R. Fukuma, B. Seymour, K. Hosomi, H. Kishima, T. Shimizu et al. 2016. Induced sensorimotor brain plasticity controls pain in phantom limb patients. Nature Communications 7: 13209. https://doi.org/10.1038/ncomms13209.

[10] Flor, H. 2008. Maladaptive plasticity, memory for pain and phantom limb pain: review and suggestions for new therapies. Expert Review of Neurotherapeutics 8(5): 809–818. https://doi.org/10.1586/14737175.8.5.809.

[11] Condés-Lara, M., F. A. Barrios, J. R. Romo, R. Rojas, P. Salgado and J. Sánchez-Cortazar. 2000. Brain somatic representation of phantom and intact limb: a fMRI study case report. European Journal of Pain (London, England): 4(3): 239–245. https://doi.org/10.1053/eujp.2000.0176.

[12] Wiech, K., H. Preissl and N. Birbaumer. 2000. Neuroimaging of chronic pain: phantom limb and musculoskeletal pain. Scandinavian Journal of Rheumatology. Supplement, 113: 13–18. https://doi.org/10.1080/030097400446571.

[13] Roux, F. E., J. A. Lotterie, E. Cassol, Y. Lazorthes, J. C. Sol and I. Berry. 2003. Cortical areas involved in virtual movement of phantom limbs: comparison with normal subjects. Neurosurgery 53(6): 1342–1353. https://doi.org/10.1227/01.neu.0000093424.71086.8f.

[14] Sposato, L., K. Yancosek, J. Lospinoso and J. Cancio. 2018. Psychosocial reactions to upper extremity limb salvage: A cross-sectional study. Journal of Hand Therapy : Official Journal of the American Society of Hand Therapists 31(4): 494–501. https://doi.org/10.1016/j.jht.2017.05.020.

[15] Jo, S. H., S. H. Kang, W. S. Seo, B. H. Koo, H. G. Kim and S. H. Yun. 2021. Psychiatric understanding and treatment of patients with amputations. Yeungnam University Journal of Medicine 38(3): 194–201. https://doi.org/10.12701/yujm.2021.00990.

[16] Bailey, A. A. and F. P. Moersch. 1941. Phantom Limb. Canadian Medical Association Journal 45(1): 37–42.

[17] Reilly, K. T. and A. Sirigu. 2008. The motor cortex and its role in phantom limb phenomena. The Neuroscientist : A Review Journal Bringing Neurobiology, Neurology and Psychiatry 14(2): 195–202. https://doi.org/10.1177/1073858407309466.

[18] Melzack, R. 1990. Phantom limbs and the concept of a neuromatrix. Trends in Neurosciences 13(3): 88–92. https://doi.org/10.1016/0166-2236(90)90179-e.

[19] Melzack, R. 1992. Phantom limbs. Scientific American 266(4): 120–126. https://doi.org/10.1038/scientificamerican0492-120.

[20] Melzack, R. and J. Katz. 2013. Pain. Wiley interdisciplinary reviews. Cognitive Science 4(1): 1–15. https://doi.org/10.1002/wcs.1201.

[21] Garraghty, P. E., E. A. LaChica and J. H. Kaas. 1991. Injury-induced reorganization of somatosensory cortex is accompanied by reductions in GABA staining. Somatosensory and Motor Research 8(4): 347–354. https://doi.org/10.3109/08990229109144757.

[22] Ramachandran, V. S. and W. Hirstein. 1998. The perception of phantom limbs. The D. O. Hebb lecture. Brain : A Journal of Neurology 121: 1603–1630. https://doi.org/10.1093/brain/121.9.1603.

[23] Ramachandran, V. S. and D. Rogers-Ramachandran. 2000. Phantom limbs and neural plasticity. Archives of Neurology 57(3): 317–320. https://doi.org/10.1001/archneur.57.3.317.

[24] Penfield, W. and E. Boldrey. 1937. Somatic motor and sensory representation in the cerebral cortex of man as studied by electrical stimulation. Brain 60: 389–443. http:// dx.doi.org/10.1093/brain/60.4.389.

[25] Pearson-Fuhrhop, K. M., J. A. Kleim and S. C. Cramer. 2009. Brain plasticity and genetic factors. Topics in Stroke Rehabilitation 16(4): 282–299. https://doi.org/10.1310/tsr1604-282.

[26] Bruurmijn, M. L. C. M., I. P. L. Pereboom, M. J. Vansteensel, M. A. H. Raemaekers and N. F. Ramsey. 2017. Preservation of hand movement representation in the sensorimotor areas of amputees. Brain: A Journal of Neurology 140(12): 3166–3178. https://doi.org/10.1093/brain/awx274.

[27] Raffin, E. 2021. The various forms of sensorimotor plasticity following limb amputation and their link with rehabilitation strategies. Revue Neurologique 177(9): 1112–1120. https://doi.org/10.1016/j.neurol.2021.09.003.

[28] Flor, H., T. Elbert, S. Knecht, C. Wienbruch, C. Pantev, N. Birbaumer et al. 1995. Phantom-limb pain as a perceptual correlate of cortical reorganization following arm amputation. Nature 375(6531): 482–484. https://doi.org/10.1038/375482a0.

[29] Karl, A., N. Birbaumer, W. Lutzenberger, L. G. Cohen and H. Flor. 2001. Reorganization of motor and somatosensory cortex in upper extremity amputees with phantom limb pain. The Journal of Neuroscience: The Official Journal of the Society for Neuroscience 21(10): 3609–3618. https://doi.org/10.1523/JNEUROSCI.21-10-03609.2001.

[30] Chen, R., B. Corwell, Z. Yaseen, M. Hallett and L. G. Cohen. 1998. Mechanisms of cortical reorganization in lower-limb amputees. The Journal of Neuroscience : The Official Journal of the Society for Neuroscience 18(9): 3443–3450. https://doi.org/10.1523/JNEUROSCI.18-09-03443.1998.

[31] Raffin, E., N. Richard, P. Giraux and K. T. Reilly. 2016. Primary motor cortex changes after amputation correlate with phantom limb pain and the ability to move the phantom limb. NeuroImage 130: 134–144. https://doi.org/10.1016/j.neuroimage.2016.01.063.

[32] Orczyk, J. J. and P. E. Garraghty. 2015. Reconciling homeostatic and use-dependent plasticity in the context of somatosensory deprivation. Neural Plasticity 2015: 290819. https://doi.org/10.1155/2015/290819.

[33] Sparling, T., L. Iyer, P. Pasquina and E. Petrus. 2024. Cortical reorganization after limb loss: bridging the gap between basic science and clinical recovery. The Journal of Neuroscience : The Official Journal of the Society for Neuroscience 44(1): e1051232024. https://doi.org/10.1523/JNEUROSCI.1051-23.2023.

[34] Howseman, A. M. and R. W. Bowtell. 1999. Functional magnetic resonance imaging: imaging techniques and contrast mechanisms. Philosophical Transactions of the Royal Society of London. Series B, Biological Sciences 354(1387): 1179–1194. https://doi.org/10.1098/rstb.1999.0473.

[35] Lotze, M., H. Flor, W. Grodd, W. Larbig and N. Birbaumer. 2001. Phantom movements and pain. An fMRI study in upper limb amputees. Brain : A Journal of Neurology 124(Pt 11): 2268–2277. https://doi.org/10.1093/brain/124.11.2268.

[36] Elbert, T., H. Flor, N. Birbaumer, S. Knecht, S. Hampson, W. Larbig et al. 1994. Extensive reorganization of the somatosensory cortex in adult humans after nervous system injury. Neuroreport 5(18): 2593–2597. https://doi.org/10.1097/00001756-199412000-00047.

[37] Kew, J. J., M. C. Ridding, J. C. Rothwell, R. E. Passingham, P. N. Leigh, S. Sooriakumaran et al. 1994. Reorganization of cortical blood flow and transcranial magnetic stimulation maps in human subjects after upper limb amputation. Journal of Neurophysiology 72(5): 2517–2524. https://doi.org/10.1152/jn.1994.72.5.2517.

[38] Proudfoot,M.,M.W.Woolrich,A.C.NobreandM.R.Turner.2014.Magnetoencephalography.Practical Neurology 14(5): 336–343. https://doi.org/10.1136/practneurol-2013-000768.

[39] Roberts, T. P. and H. A. Rowley. 1997. Mapping of the sensorimotor cortex: functional MR and magnetic source imaging. AJNR. American Journal of Neuroradiology 18(5): 871–880.

[40] Horgan, O. and M. MacLachlan. 2004. Psychosocial adjustment to lower-limb amputation: a review. Disability and Rehabilitation 26(14-15): 837–850. https://doi.org/10.1080/0963828041000 1708869.

[41] Shue, S., Y. Wu-Fienberg and K. J. Chepla. 2021. Psychiatric disease after isolated traumatic upper extremity amputation. Journal of Hand and Microsurgery 13(2): 75–80. https://doi.org/10.1055/s-0040-1701156.

[42] Coffey, L., F. O'Keeffe, P. Gallagher, D. Desmond and R. Lombard-Vance. 2012. Cognitive functioning in persons with lower limb amputations: a review. Disability and Rehabilitation 34(23): 1950–1964. https://doi.org/10.3109/09638288.2012.667190.

[43] O'Neill, B. F. and J. J. Evans. 2009. Memory and executive function predict mobility rehabilitation outcome after lower-limb amputation. Disability and Rehabilitation 31(13): 1083–1091. https://doi.org/10.1080/09638280802509579.

[44] Resnik, L. J., P. Ni, M. L. Borgia and M. A. Clark. 2022. A psychosocial adjustment measure for persons with upper limb amputation. Canadian Prosthetics and Orthotics Journal 5(1): 37873. https://doi.org/10.33137/cpoj.v5i1.37873.

[45] Desteli, E. E., Y. İmren, M. Erdoğan, G. Sarısoy and S. Coşgun. 2014. Comparison of upper limb amputees and lower limb amputees: a psychosocial perspective. European Journal of Trauma and Emergency Surgery : Official Publication of the European Trauma Society 40(6): 735–739. https://doi.org/10.1007/s00068-014-0418-3.

[46] Saradjian, A., A. R. Thompson and D. Datta. 2008. The experience of men using an upper limb prosthesis following amputation: positive coping and minimizing feeling different. Disability and Rehabilitation 30(11): 871–883. https://doi.org/10.1080/09638280701427386.

[47] Desmond, D. M. 2007. Coping, affective distress, and psychosocial adjustment among people with traumatic upper limb amputations. Journal of Psychosomatic Research 62(1): 15–21. https://doi.org/10.1016/j.jpsychores.2006.07.027.

Index

For Product Safety Concerns and Information please contact our EU representative GPSR@taylorandfrancis.com Taylor & Francis Verlag GmbH, Kaufingerstraße 24, 80331 München, Germany